PHARMACOLOGY - RESEARCH, SAFETY TESTING AND REGULATION

ADENOSINE RECEPTORS

PHARMACOLOGY, FUNCTIONS AND THERAPEUTIC ASPECTS

PHARMACOLOGY - RESEARCH, SAFETY TESTING AND REGULATION

Additional books in this series can be found on Nova's website under the Series tab.

Additional e-books in this series can be found on Nova's website under the e-book tab.

PHARMACOLOGY - RESEARCH, SAFETY TESTING AND REGULATION

ADENOSINE RECEPTORS

PHARMACOLOGY, FUNCTIONS AND THERAPEUTIC ASPECTS

KASANDRA WARRICK
EDITOR

NOVA BIOMEDICAL
New York

Library of Congress Cataloging-in-Publication Data

ISBN: 978-1-63463-454-0

Library of Congress Control Number: 2014954522

Published by Nova Science Publishers, Inc. † New York

Contents

Preface

Adenosine is a ubiquitous adenine base purine that mediates diverse biological and physiological processes in the cardiovascular system, both in cardiac and vascular tissues. Adenosine was reported to be involved in smooth muscle contraction, neurotransmission/neuromodulation, immune response and inflammation, platelet aggregation and modulation of cardiac function. Adenosine availability, in the extracellular space, depends on its release from a variety of cell types, as a result of adenosine triphosphate (ATP) enzymatic breakdown of both intra- and extracellular adenine nucleotides and intracellular S-adenosylhomocysteine. This book discusses the pharmacology, functions and therapeutic aspects of adenosine receptors.

Chapter I – Organ transplantation unavoidably entails ischemia-reperfusion injury (IRI) resulting in graft dysfunction and increased morbidity and mortality. Ischemia, the loss of blood supply to an organ or tissue, results in oxygen deprivation and cellular injury. Reperfusion, the restoration of blood flow and thus oxygen to ischemic tissue, further exacerbates injury and has both short- and long-term effects on organ function. IRI is a rapid inflammatory response involving oxidative stress, cytokine production, innate immune cell activation, activation of adhesion molecules and leukocyte infiltration. Adenosine is a nucleoside that is locally released in response to cellular stress such as IRI and largely serves as an endogenous mechanism to combat excessive inflammation. Adenosine signals through four G protein-coupled receptor subtypes: A1R, A2AR, A2BR, and A3R. Because adenosine has a very short half-life and is rapidly metabolized, its therapeutic use is limited. Thus, potent and selective adenosine receptor agonists have been synthesized and utilized in many experimental studies to dissect the roles of each receptor in organ injury. A2AR activation is well documented to potently attenuate IRI and other inflammatory conditions in lung, heart, liver and kidney, and the anti-inflammatory effects of A2AR agonists on various leukocytes have been demonstrated. A2AR signaling is classically linked to stimulatory G proteins resulting in up-regulation of cAMP production and transcriptional activation of cAMP-regulated genes. This chapter examines the role of A2AR activation in IRI and the anti-inflammatory and immunosuppressive potential of pharmacologic A2AR agonists. Currently no therapeutic strategies are clinically available that directly targets IRI, and A2AR agonists represent a promising avenue for the prevention of IRI after transplantation. This novel therapeutic strategy is reviewed as a foundational discussion for future research initiatives in the prevention of IRI.

Chapter II – Adenosine is a ubiquitous adenine base purine that mediates diverse biological and physiological processes in the cardiovascular system, both in cardiac and vascular tissues. Adenosine was reported to be involved in smooth muscle contraction, neurotransmission/neuromodulation, immune response and inflammation, platelet aggregation and modulation of cardiac function. Adenosine availability, in the extracellular space, depends on its release from a variety of cell types, as a result of adenosine triphosphate (ATP) enzymatic breakdown of both intra- and extracellular adenine nucleotides and intracellular S-adenosylhomocysteine. Other important source of extracellular adenosine lies on the activity of nucleoside transporters, both equilibrative (ENT) and concentrative (CNT), particularly ENT1, ENT4 and CNT2. As such adenosine can reach concentrations able to activate adenosine receptors. Four adenosine receptor subtypes have been identified, A_1, A_{2A}, A_{2B} and A_3 receptors, and were found to trigger different signalling pathways and to present differential affinity requirements. Correlation between the ENT activity and adenosine receptor activation is also discussed focusing on the impact that changes or disruption of the interplay between these effectors may cause in the vascular physiology, namely by contributing to disease. All adenosine receptor subtypes have been identified in the vascular wall of many different arteries and veins. A brief overview of the current knowledge and of the authors' research work will be integrated considering important aspects of the adenosine-vascular interaction with particular emphasis on adenosine receptor subtypes pharmacology in the three vascular layers, intima, media and adventitia, namely by explaining their individual contribution for vasoconstriction/vasodilation. Integration of these knowledge with the morphological features of arteries/veins, namely by evidencing how innervation may vary from dense to sparse (depending on vessel localization) or how smooth muscle cell layers of the media varies (depending on vessel type and/or localization) will also be discussed, clarifying the contribution of adenosine receptors to peripheral vascular resistance. In the last decade several new adenosine receptor ligands have been identified as having promising therapeutic activities but during clinical trials some of them have been discontinued due to side effects. Despite of this fact, a renewed interest in this research area is emerging and several new adenosine receptor ligands have been developed and are undergoing clinical trials for a broad range of indications such as neurodegenerative disorders, inflammation, wound healing, cardiac imaging, etc. In summary, in this chapter an overview of adenosine/adenosine receptors physiology and pharmacology in the vasculature is presented guiding the reader to new and putative therapeutic applications.

Chapter III – The role of adenosine and adenosine receptors in tumourigenesis has been largely investigated. Studies suggest a contradictory role of adenosine in viability of normal and cancer cells, ascribed to the stimulation of adenosine receptors subtypes, named A_1, A_{2A}, A_{2B}, and A_3, which couple to different signal transduction pathways and can be co-expressed in the same cell. Several studies have suggested a crucial role for A_3 adenosine receptors on cancer cells' cytotoxicity, improving, by opposition, normal cells proliferation. This dual effect, coupled with the fact that these receptors are overexpressed in several tumours and expressed at low levels in normal tissues, suggested A_3 agonists as promising cancer therapeutic strategies. However, time has proved this to be an oversimplified view. In fact, activation of adenosine A_3 receptors can lead both to malignant cell proliferation and death, depending on agonist's concentration. Moreover, A_3 agonists can mediate cytotoxicity independently of adenosine A_3 receptor activation. Melanoma is considered one of the most aggressive malignant tumours, due to its high metastatic potential and resistance to current treatment modalities. Patients with metastatic melanoma present poor prognosis, with an overall survival of 8-18 months. Melanoma incidence is increasing faster than any other solid tumour accounting for a global annual incidence of about 160 thousands new cases. This chapter intends to integrate the authors' research work and the current knowledge concerning the role of adenosine/adenosine receptors in the complex interplay of proliferation, cell death mechanisms and metastatic progression of melanoma. The authors' group has shown adenosine receptor subtypes expression, on different metastatic melanoma cell lines. Their activation did not affect cell proliferation of human A375 and mouse K1735-M2 cells. However, proliferation of human C32 cells was increased, after activation of A_3 adenosine receptor, by micromolar concentrations of adenosine or nanomolar concentrations of the selective A_3 agonist, Cl-IB-

MECA. By opposition, Cl-IB-MECA, at micromolar concentrations, caused marked cytotoxicity, through A_3-independent mechanisms. A major role for adenosine A_3 receptors in C32 cell proliferation induced by an adenosine metabolite, inosine, through PLC-PKC-MEK1/2-ERK1/2 pathway activation was demonstrated. Inosine also enhanced proliferation of A375 cells and stimulated metastatic processes evaluated in both cell lines (migration, adhesion, invasion and colony-formation), through adenosine A_3 receptor activation. The combination Cl-IB-MECA and paclitaxel was investigated and found to induce multiple mechanisms of cell death and inhibit metastatic processes. This combination emerges, therefore, as a promising new therapeutic strategy for human metastatic melanoma, potentially overcoming chemotherapy multiresistance and improving patient prognosis.

Chapter IV – There is a considerable bulk of evidence for the relevance of ATP in the regulation of bladder function in health and in disease conditions. However the actions of its metabolite, adenosine, on bladder tissue only recently are being unveiled. All four subtypes of adenosine receptors (A_1, $A_{2a}A_{2A}$, A_{2B} and A_3) have been identified in the bladder of experimental animals by RT-PCR and Western blot analysis; these receptors are differently distributed among the urothelium and detrusor layers. Adenosine receptors are also present in the human bladder where they can modulate filling sensations and voiding. While adenosine inhibits nerve-evoked acetylcholine release from cholinergic nerve efferents, via high affinity A_1 receptors, high concentrations of the nucleoside are required to reduce detrusor contractions induced by carbachol, ATP and potassium depolarizations. Characterization of the adenosine receptor involved in detrusor smooth muscle relaxation is still a matter of debate. Besides the major effects of the nucleoside on bladder voiding, adenosine also influences the filling phase of the micturition reflex by controlling the release of ATP from urothelial cells and the firing of sensory nerve afferents, via A_1 receptors activation. Differences in the kinetics of ATP metabolism by ectonucleotidases and adenosine biosynthesis between luminal and serosal sides of the urothelium may explain the dominant role of adenosine in the suburothelium and detrusor smooth muscle layers. Thus, the adenosine signaling pathway is now being explored as a potential therapeutic target for human bladder disorders, including inflammatory diseases, overactivity and outflow obstruction.

Chapter V – Methotrexate (MTX) is an anti-tumor medicine classified into the anti-folate drugs. MTX suppresses DNA synthesis through inhibiting dihydrofolate reductase (DHF) of *de novo* nucleic acid pathway, which suppresses proliferation of malignant tumors. MTX is also utilized as for anti-rheumatic medicine as the lower doses of MTX are able to suppress effectively the inflammation of the synovial tissues in RA patients. In the authors' previous study, plasma adenosine level was markedly increased during induction of inflammation in rats with adjuvant-induced arthritis (arthritic rats). MTX-induced suppression of osteoclastogenesis was canceled by the addition of adenosine *in vitro* as well as *in vivo*. Deoxyadenosine (dAdo) is known to regulate proliferation of immune cells and partly share metabolic pathways with adenosine. Adenosine deaminase (ADA) is a key enzyme which metabolize both of adenosine and dAdo to produce inosine and deoxyinosine perpetually. Here the authors examined the regulatory role of dAdo in MTX-induced suppression of inflammatory bone destruction and investigated a possible involvement of ADA as a target molecule of MTX. In rats with adjuvant-induced arthritis, in whole bone marrow cultures for evaluating osteoclastogenesis, MTX markedly suppressed osteoclastogenesis and dAdo completely canceled its suppression. This cancellation was partially blocked by caffeine, antagonist for the adenosine receptors, A_1AR, $A_{2a}AR$ and A_3AR. Semi-quantitative RT-PCR showed that MTX suppressed expression of RANKL without affecting osteoprotegerin (OPG) expression. Addition of dAdo clearly recovered the expression of RANKL and slightly suppressed OPG expression, which result in augment in osteoclastogenesis. In arthritic rats, MTX-induced suppression of inflammatory bone destruction was completely cancelled by the injection of dAdo. A marked induction of ADA was apparent in bone marrow cells observed in the bone destruction sites around ankle joints in arthritic rats. MTX strikingly suppressed expression of ADA in the corresponding area. Immunoblot analysis also showed a clear induction of ADA protein in arthritic rats and a marked suppression of ADA protein when arthritic rats were treated with MTX. In bone marrow cultures for evaluating osteoclastogenesis, expression of ADA mRNA was markedly suppressed by MTX and dAdo partially cancelled its inhibition. These observations suggest that ADA is involved in the MTX-mediated inhibition of inflammatory bone destruction regulated by adenosine and dAdo. Suppression of ADA by MTX may contribute to the stability of dAdo. A list of abbreviation: AA, adjuvant-induced arthritis; ADA, adenosine deaminase; AR, adenosine receptor; CFA, complete Freund adjuvant; dAdo, deoxyadenosine; MNC, multinucleated cells; MTX, methotrexate; RA, rheumatoid arthritis; RANK, receptor activator NF-kB ligand, TRAP, tartrate acid-resistant phosphatase.

In: Adenosine Receptors
Editor: Kasandra Warrick
ISBN: 978-1-63463-454-0

Chapter I

Anti-Inflammatory and Immunosuppressive Actions of Adenosine A2A Receptor in Ischemia-Reperfusion Injury after Organ Transplantation

Ashish K. Sharma[1], *Gilbert R. Kinsey*[2], *Preeti Chhabra*[1], *Kenneth L. Brayman*[1], *Brent A. French*[3], *David K. Glover*[2] *and Victor E. Laubach*[1*]

[1]Department of Surgery, University of Virginia, Charlottesville, VA, US
[2]Department of Medicine, University of Virginia, Charlottesville, VA, US
[3]Department of Biomedical Engineering, University of Virginia, Charlottesville, VA, US

* Corresponding author: Department of Surgery, University of Virginia, P.O. Box 801359, Charlottesville, VA 22908, Phone: 434-924-2927, Email: laubach@virginia.edu.

Abstract

Organ transplantation unavoidably entails ischemia-reperfusion injury (IRI) resulting in graft dysfunction and increased morbidity and mortality. Ischemia, the loss of blood supply to an organ or tissue, results in oxygen deprivation and cellular injury. Reperfusion, the restoration of blood flow and thus oxygen to ischemic tissue, further exacerbates injury and has both short- and long-term effects on organ function. IRI is a rapid inflammatory response involving oxidative stress, cytokine production, innate immune cell activation, activation of adhesion molecules and leukocyte infiltration. Adenosine is a nucleoside that is locally released in response to cellular stress such as IRI and largely serves as an endogenous mechanism to combat excessive inflammation. Adenosine signals through four G protein-coupled receptor subtypes: A1R, A2AR, A2BR, and A3R. Because adenosine has a very short half-life and is rapidly metabolized, its therapeutic use is limited. Thus, potent and selective adenosine receptor agonists have been synthesized and utilized in many experimental studies to dissect the roles of each receptor in organ injury. A2AR activation is well documented to potently attenuate IRI and other inflammatory conditions in lung, heart, liver and kidney, and the anti-inflammatory effects of A2AR agonists on various leukocytes have been demonstrated. A2AR signaling is classically linked to stimulatory G proteins resulting in up-regulation of cAMP production and transcriptional activation of cAMP-regulated genes. This chapter examines the role of A2AR activation in IRI and the anti-inflammatory and immunosuppressive potential of pharmacologic A2AR agonists. Currently no therapeutic strategies are clinically available that directly targets IRI, and A2AR agonists represent a promising avenue for the prevention of IRI after transplantation. This novel therapeutic strategy is reviewed as a foundational discussion for future research initiatives in the prevention of IRI.

Introduction

Ischemia is a lack of blood supply to an organ or tissue, resulting in oxygen deprivation of cells, which is unavoidable in the case of organ transplantation. Without restoration of oxygenated blood, ischemic tissues soon undergo necrosis and die. Thus, restoration of blood flow, or reperfusion, is the definitive treatment for ischemia. Although ischemia itself is a serious condition, reperfusion of ischemic tissue upon transplantation paradoxically imparts further tissue injury called ischemia-reperfusion injury (IRI). IRI is a rapid and complex cascade of inflammatory events that entails oxidative stress, innate immune responses, leukocyte infiltration, cytokine and chemokine release and cell/tissue damage. Under severe conditions IRI imposes a considerable threat to graft and recipient survival and thus contributes to significant morbidity and mortality after transplantation.

Currently no clinical therapies are available to prevent IRI, and treatment strategies are limited to supportive therapy. Despite our current understanding of the causes and effects of IRI, IRI precedes clinical symptoms, and once apparent, it is often too late for intervention. Therefore preventative therapeutic strategies offer the best hope for reducing the number of transplant patients who suffer or die from IRI. The primary determinants for the severity of IRI are quality of the donor allograft and nature of recipient pathophysiology. Strategies commonly used to minimize the severity of IRI include 1) optimizing selection, matching and management of donors and recipients preoperatively, 2) improving lung preservation methods, and 3) improving lung implantation and reperfusion techniques. The standard methods for organ preservation involve flushing of the donor organ with a cold crystalloid preservation solution and cold static storage. Cold storage during the preservation period limits metabolic activity, vasospasm, and thrombosis. Reimplantation into the recipient restores warm perfusion to the allograft, initiating the characteristic IRI inflammatory cascade.

1.1. Adenosine and Adenosine Receptors

In response to cellular stress such as IRI, extracellular adenosine levels rapidly increase. Once in the extracellular space, adenosine can act as a signaling molecule in an autocrine or paracrine manner to elicit a range of tissue responses that can generally be viewed as organ protective. Extracellular adenosine largely accumulates as a result of ectonucleotidase breakdown of precursor nucleotides, such as ATP, ADP and AMP, which are released by many cell types including endothelial cells, inflammatory cells and epithelial cells (Figure 1). CD39 ectonucleotidase converts extracellular ATP (or ADP) to AMP whereupon CD73 ectonucleotidase converts AMP to adenosine. Adenosine is also transported between extracellular and intracellular stores by both equilibrative nucleoside transporters (ENTs), which mediate the equilibrium of adenosine concentrations by facilitated transport, and concentrative nucleoside transporters (CNTs), which allow for the preservation of adenosine concentration gradients across cell membranes [1-3]. Adenosine is rapidly metabolized by extracellular or intracellular adenosine deaminase, resulting in the production of inosine, or phosphorylated by intracellular adenosine kinase to form AMP [4, 5]. Adenosine concentrations vary greatly by tissue and physiological state with the levels increasing multifold during metabolically demanding conditions of oxidative stress, hypoxia, exercise or inflammation [6-10]. In particular, hypoxia has been shown to inhibit adenosine kinase [11].

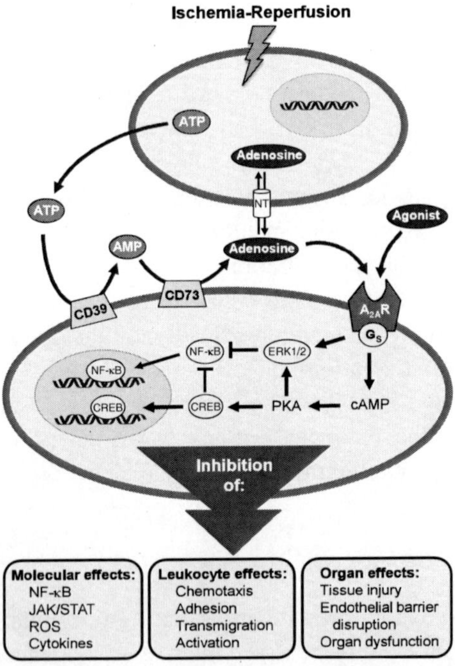

Figure 1. Adenosine A2A receptor ($A_{2A}R$) activation attenuates inflammation after ischemia-reperfusion. Ischemia-reperfusion induces localized accumulation of extracellular ATP via multiple mechanisms (e.g. injury, necrosis and nucleotide permeable channels). CD39 ectonucleotidase converts extracellular ATP to AMP, and CD73 ectonucleotidase converts AMP to adenosine. Adenosine can also be released through nucleoside transporters (NT). Classically, but not necessarily, adenosine or agonist binding to $A_{2A}R$, a G_s protein-coupled receptor, stimulates the production of cAMP and subsequent activation of protein kinase A (PKA). PKA then activates cAMP responsive element binding protein (CREB) to induce the transcription of genes containing CREB-responsive elements. $A_{2A}R$ can also activate ERK1/2 via PKA dependent or independent pathways. Other cAMP-independent pathways of $A_{2A}R$ activation include PI3K/Akt signaling (not shown). $A_{2A}R$ signaling results in the attenuation of inflammatory pathways (e.g. NF-κB activation and proinflammatory cytokine production) and leukocyte activation/infiltration, which protect organs from ischemia-reperfusion-induced tissue injury, endothelial barrier disruption and organ dysfunction. ROS, reactive oxygen species.

Extracellular adenosine signals through the binding and activation of G-protein coupled cell surface receptors known as adenosine receptors, of which there are four subtypes: A_1R, $A_{2A}R$, $A_{2B}R$ and A_3R [12]. These receptors reside on a wide variety of cell types, and depending on the relative expression pattern of adenosine receptors on different cell and tissue types, activation and signaling result in different biologic functions. The A_1R and A_3R are coupled to G_i and thereby cause an inhibition of the cAMP pathway whereas the $A_{2A}R$ and $A_{2B}R$ are coupled to G_s, which activates adenylate cyclase and consequently elevates cAMP levels [13. As these two pathways are often activated simultaneously, the end result is an elevation in intracellular calcium. A_1R, $A_{2A}R$ and A_3R are high-affinity receptors activated by low concentrations (10 nM–1 μM) of adenosine, which are present under physiological conditions. In contrast, the $A_{2B}R$ is a low-affinity receptor requiring higher concentrations of adenosine (10 μM) for receptor activation [14]. These levels of adenosine are typically seen in pathologic disturbances, such as during periods of hypoxia, ischemia and other cellular distress [15]. Adenosine has a very short half-life in the extracellular space due in part to its rapid uptake into the intracellular space by nucleoside transporters. Adenosine receptor signaling can also activate other pathways involving phospholipase C, mitogen activated protein kinase (MAPK), phosphatidylinositol 3-kinase (PI3K), or Akt pathways to influence gene expression.

1.2. Anti-Inflammatory Mechanisms of $A_{2A}R$ Activation

It well known that adenosine has protective actions for many organ systems, such as the heart, brain, kidney, skeletal muscle and adipose tissue [16]. This evidence led Newby et al. to hypothesize that adenosine acts as a "retaliatory metabolite" that protects the cell against excessive external stimulation [17]. Many preclinical studies provide evidence that adenosine has profound anti-inflammatory effects via $A_{2A}R$ signaling. Expression of $A_{2A}R$ is widely distributed but is predominantly expressed on immune cells (e.g. lymphocytes, neutrophils and macrophages), and its activation by adenosine or adenosine analogues ($A_{2A}R$ agonists) results in a wide range of anti-inflammatory and immunosuppressive responses. Gene expression of $A_{2A}R$ is highly responsive to alterations in the extracellular environment. In this regard, $A_{2A}R$ expression is very sensitive to changes in the concentrations of exogenous and endogenous factors involved in inflammation. For example, exposure of macrophages to LPS induces a dramatic increase in the expression of $A_{2A}R$ mRNA and protein, effects that are brought about through activation of the transcription factor NF-κB [18].

IRI leads to immune cell activation and a robust production of proinflammatory cytokines and chemokines as well as generation of reactive oxygen species in various organ systems. The potent anti-inflammatory actions of $A_{2A}R$ agonists have been well established, and $A_{2A}R$ activation suppresses activation of immune cells such as T cells, monocytes, macrophages, neutrophils and dendritic cells [19]. Elevated cAMP and activation of protein kinase A (PKA) due to $A_{2A}R$ signaling leads to phosphorylation and activation of the transcription factor cAMP responsive element binding protein (CREB) to modulate inflammatory gene expression (Figure 1). In addition to cAMP-dependent pathways, the signaling mechanisms of $A_{2A}R$ have also been shown to be cAMP-independent, especially in neutrophils [20-22]. Largely through the implementation of selective $A_{2A}R$ agonists, *in vitro* studies provide robust evidence supporting anti-inflammatory actions of $A_{2A}R$ signaling on specific cell types, and many *in vivo* studies also exemplify the anti-inflammatory and tissue-protective effects elicited by $A_{2A}R$ agonists in numerous animal models of inflammatory disease [23]. For example, $A_{2A}R$ agonists have been shown to decrease 1) TNF-α production by monocytes and macrophages [24], 2) neutrophil-endothelial cell adherence [25] and 3) neutrophil, platelet and endothelial cell activation [26-28]. These studies and others have established pharmacologic targeting of $A_{2A}R$ activation as a promising strategy to prevent or attenuate acute inflammatory conditions including post-transplant IRI.

The signaling pathways activated downstream of $A_{2A}R$ vary with the type of cell and the signaling machinery that the cell possesses. $A_{2A}R$ signaling is able to dampen activation of two major proinflammatory pathways: the NF-κB and the JAK/STAT pathways (Figure 1). The mechanisms behind the inhibitory effects of $A_{2A}R$ activation on the NF-κB pathway differ between cell types. One mechanism that may prevent activation of NF-κB is through inhibition of the phosphorylation and degradation of IκBα, an inhibitor of NF-κB nuclear translocation when dephosphorylated. Nadeem et al. demonstrated that $A_{2A}R$ knockout mice had increased phosphorylation of IκBα resulting in increased NF-κB-dependent transcription of pro-inflammatory genes [29]. Supporting this data, splenocytes from $A_{2A}R$ knockout mice challenged with LPS were shown to have increased mRNA levels of NF-κB mediated pro-inflammatory cytokines, and wild-type animals challenged with LPS and pre-treated with $A_{2A}R$ agonists showed inhibition of LPS-induced cytokine mRNA expression [30]. Another generalized mechanism by which $A_{2A}R$ suppresses inflammation in a number of cell types is through inhibition of the JAK/STAT pathway [31]. This pathway transduces signals from receptors for hematopoietic cytokines including the IL-6 family.

Understanding of the potential target cell mediators involved in the anti-inflammatory mechanisms of $A_{2A}R$ agonists is derived mainly from studies involving IRI. Numerous studies have demonstrated that much of the anti-inflammatory actions of $A_{2A}R$ agonists during IRI are due to effects on CD4+ T cells [32-35]. Recent studies have identified invariant natural killer T (NKT) cells, a subset of CD4+ T cells, as the target cell of the anti-inflammatory effects of $A_{2A}R$ agonists [33, 36]. In addition to T cells, macrophages and dendritic cells have also been shown to be targets of $A_{2A}R$ agonists. The expression and function of $A_{2A}Rs$ on macrophages can be upregulated under inflammatory conditions. Furthermore, increased $A_{2A}R$ expression leads to inhibition of proinflammatory IL-12 secretion and upregulation of anti-inflammatory IL-10 secretion by macrophages [37]. Similarly, $A_{2A}R$ agonist treatment has been shown to modulate NKT-dendritic cell interactions to downregulate IFN-γ expression and upregulate IL-10 expression in kidney IRI [38]. Taken together, these data suggest that regulation of cells of the innate immune system by $A_{2A}R$ activation is critical for the downregulation and polarization of immune responses in the setting of IRI.

1.3. Role of $A_{2A}R$ in Lung IRI

$A_{2A}Rs$ are widely expressed in the lung where they exert a wide range of physiological effects, most of which are anti-inflammatory and reparative in nature [19]. Lung IRI is characterized by nonspecific alveolar damage, lung edema, and hypoxemia occurring within 72 hours after transplantation. IRI is the leading cause of morbidity and mortality in the early post-transplant period, with rates as high as 41% [39]. The 30-day mortality of patients with IRI is 40% compared to 7% in patients without [40]. Patients with lung IRI often demonstrate primary graft dysfunction (PGD) and require prolonged ventilation, longer hospital stay and have a higher risk of multi-organ failure [41, 42]. IRI is also a risk factor for late graft rejection (bronchiolitis obliterans), the major cause of mortality beyond 1 year of transplant [43].

Studies have shown that $A_{2A}R$ activation produces anti-inflammatory responses, prevents leukocyte adhesion, and inhibits release of reactive oxygen species after IR [35, 36, 44]. These studies have demonstrated the ability of $A_{2A}R$ agonists to attenuate the production of proinflammatory cytokines such as TNFα, IL-17, IL-6, CCL5, CCL2, CXCL1 and CXCL2. The specific activation of $A_{2A}R$ during reperfusion attenuates lung injury and dysfunction in both mouse and rabbit models of IRI as well as in a clinically relevant, pig lung transplant model [35, 45-47]. Here, $A_{2A}R$ agonists were shown to attenuate pulmonary dysfunction and lung injury as evidenced by reduced pulmonary edema, neutrophil infiltration, proinflammatory cytokine production, and microvascular leak after lung transplantation.

In recent studies, it has been shown that treatment with ATL313, a selective $A_{2A}R$ agonist, prevented lung IRI in wild type mice but had no effect in $A_{2A}R$ knockout mice [46]. The protective effect of $A_{2A}R$ agonist was abolished by ZM241385, an $A_{2A}R$ antagonist, in a rabbit lung IRI model [48]. Rivo et al. demonstrated that the $A_{2A}R$-mediated protection is associated with decreased apoptosis and involves ERK1/2 activation and modulation of Bcl-2 and Bax expression [49]. Moreover, in a pig lung transplant model, treatment with ATL146e, an $A_{2A}R$ agonist, attenuated lung injury and inflammation thereby preserving pulmonary function after transplantation [50]. Furthermore, the enhancement of donor lung preservation and function by $A_{2A}R$ agonist treatment during *ex vivo* lung perfusion (EVLP) was shown to be superior as compared to EVLP alone [51]. Taken together, these studies underline the importance of $A_{2A}R$ agonists as a promising therapeutic strategy for the prevention of IRI and PGD in lung transplant patients.

In addition to animal models of asthma and COPD [23], $A_{2A}R$ activation has also been shown to be important in protection against acute lung injury and acute respiratory distress syndrome. For example, $A_{2A}R$ activation using an agonist, CGS21680, ameliorated lung injury and neutrophil sequestration caused by trauma in combination with hemorrhagic shock [52]. In addition, CGS21680 was shown to attenuate ventilator-induced lung injury in rats as evidenced by a decrease in edema, neutrophil infiltration and inflammatory cytokine production [53]. Reutershan et al. used an $A_{2A}R$ agonist to demonstrate that activation of $A_{2A}Rs$ on myeloid cells attenuates recruitment and adhesion of neutrophils to pulmonary endothelial cells following LPS-induced lung injury [54]. The fact that selective $A_{2A}R$ agonists are currently being tested in clinical trials for other diseases such as sickle cell crisis [55] makes $A_{2A}R$ an attractive target for therapeutic intervention in human acute lung injury and acute respiratory distress syndrome including IRI. As with other immunomodulatory strategies, possible adverse effects that are associated with the biologic function of $A_{2A}R$, such as inhibition of platelet aggregation or vasodilation, require attention [56]. The development of agents that allow targeted delivery to the lung via aerosol may be one concept to overcome these adverse effects.

1.4. Role of $A_{2A}R$ in Kidney IRI

IR of the kidney is a major contributor to acute kidney injury (AKI; previously called acute renal failure) and delayed graft function in renal allografts [57, 58]. AKI affects more than 5% of all hospitalized patients and the incidence increases to 40% or more in the critically ill [59, 60]. During AKI, kidney function is impaired resulting in reduced clearance of fluid, electrolytes, drugs and metabolic waste products. In some cases AKI resolves over several days; however in severe cases hemodialysis is necessary to prevent fluid overload, metabolic acidosis or life threatening hyperkalemia. It has recently been appreciated that episodes of AKI predispose individuals to chronic and end stage renal disease later in life [61, 62]. The pathogenesis of AKI involves direct ischemic damage to renal tubular epithelial cells, intra-renal vasoconstriction, endothelial cell injury and dysfunction, and accumulation/activation of immune cells during reperfusion that compounds the direct ischemic damage to kidney cells [63].

Because all four adenosine receptors are expressed in different cells of the kidney [64], the role of adenosine in kidney IRI has been investigated extensively in an effort to identify novel therapeutic agents. Animal studies have revealed that each one of the adenosine receptors plays an important role in kidney IRI. For example, A_1R, $A_{2A}R$ and $A_{2B}R$ activation are beneficial during kidney IRI [65-67], whereas activation of A_3R receptor promotes renal damage after ischemia [68]. The first study of the independent role of $A_{2A}R$ in kidney IRI employed a potent and selective $A_{2A}R$ agonist called DWH-146e (or ATL146e) that was infused continuously for 5 hours prior to ischemia which reduced kidney injury and dysfunction measured at 24 and 48 hours of reperfusion in rats [67]. The question of whether endogenous adenosine acts on $A_{2A}Rs$ in a beneficial way was investigated by comparing the response to kidney IR in $A_{2A}R$ knockout (KO) versus wild-type (WT) mice, in the absence of $A_{2A}R$ agonist treatment. The $A_{2A}R$ KO mice had significantly worse renal function at 24 hours of reperfusion following equivalent kidney ischemic times [32]. These studies demonstrate that activation of the $A_{2A}R$ by endogenous adenosine and by exogenous pharmaceutical agonists reduces kidney IRI.

The protective role of $A_{2A}R$ in kidney IRI could be due to several mechanisms including the previously mentioned anti-inflammatory actions of

$A_{2A}R$ activation on immune cells. Alternatively, adenosine is known to influence renal hemodynamics [69], and infusion of the $A_{2A}R$ selective agonist CGS21680 transiently increases renal blood flow [70]. Furthermore, CGS21680 treatment of isolated renal outer medullary descending vasa recta (the blood vessels that supply oxygen and nutrients to the highly IR-susceptible outer medulla of the kidney) counteracts the vasoconstriction induced by angiotensin II exposure [71]. Lastly, initial studies into the protective role of $A_{2A}R$ agonists in kidney IRI demonstrated that during reperfusion kidney vascular endothelial cell expression of P-selectin and intercellular adhesion molecule 1 (ICAM-1) was significantly reduced by $A_{2A}R$ activation [72]. Thus, given the expression pattern of the $A_{2A}R$ and the results of $A_{2A}R$ activation in different cell types described above, the protection afforded by $A_{2A}R$ activation in kidney IRI may be due to actions on vascular smooth muscle cells, endothelial cells, immune cells or some combination thereof.

To investigate the relative contribution of $A_{2A}Rs$ on immune vs. non-immune cells, Okusa et al. generated bone marrow chimeric mice in which WT mice were reconstituted with bone marrow isolated from WT or $A_{2A}R$ KO mice to create $A_{2A}R$ KO bone marrow → WT mouse chimeras (mice lacking $A_{2A}R$ only in the immune cell compartment) and WT bone marrow → WT mouse chimeras [32]. Kidney IR resulted in significant renal dysfunction in both types of chimeric mice, and $A_{2A}R$ agonist treatment significantly reduced injury only in the WT bone marrow → WT mouse control chimeras, demonstrating that the protection from kidney IRI is mediated through effects of $A_{2A}Rs$ on immune (bone marrow-derived) cells.

Subsequent studies have identified several types of immune cells in which $A_{2A}R$ signaling promotes protection from kidney IRI. For example, Rag-1 KO mice (which lack T and B cells) are unable to be protected from kidney IRI by $A_{2A}R$ agonists; however, reconstitution of Rag-1 KO mice with isolated WT CD4+ T cells restored $A_{2A}R$ agonist-mediated protection [73]. This finding suggests that activation of $A_{2A}Rs$ on one or more type(s) of CD4+ T cells promotes resistance to kidney IRI. One subset of $CD4^+$ T cells known to protect the kidney from IRI is the regulatory T cell (Treg) [74]. Tregs are anti-inflammatory lymphocytes identified by expression of the FoxP3 transcription factor [75-77]. Interestingly, extracellular adenosine generation (through CD39 and CD73 expressed on the surface of Tregs) is one of the many ways Tregs inhibit inflammation [78]. In a mouse model of kidney IR, depletion of Tregs prior to IR exacerbates injury and dysfunction during reperfusion [74]. On the other hand, adoptive transfer of isolated WT Tregs markedly reduces kidney injury caused by IR [74, 79]. However, adoptive transfer of CD73 KO or $A_{2A}R$ KO Tregs offers no protection from kidney IRI, suggesting that the ability to generate and respond to adenosine via $A_{2A}R$ activation is required for Treg-mediated protection from kidney IRI [80]. *Ex vivo* incubation of WT or CD73 KO Tregs with a selective $A_{2A}R$ agonist (with subsequent washing) prior to adoptive transfer, enhanced (for WT Tregs) or partially restored (for CD73 KO Tregs) the ability of Tregs to protect the kidney from IRI [80]. Taken together these results suggest that Tregs are one of the targets of systemically administered $A_{2A}R$ agonists that promote protection from kidney IRI.

Another potential CD4+ target of $A_{2A}R$ agonists are NKT cells which are known to modulate renal IRI through rapid production of the proinflammatory cytokine IFN-γ [81, 82]. Systemic ATL146e treatment reduces the accumulation of IFN-γ-producing NKT cells in the kidney during reperfusion [38]. Overall, CD4+ T cells are required for $A_{2A}R$-mediated protection from kidney IRI, and Tregs and NKT cells represent two critical targets in this pathway.

Another type of immune cell that is centrally involved in the pathogenesis of kidney IRI and that expresses the $A_{2A}R$ is the dendritic cell (DC) [38, 83]. Kidney DCs are specialized antigen presenting cells that express CD11c, F4/80 and MHC II and are the most prominent immune cell type in the normal kidney [84, 85]. Upon kidney IR, DCs produce pro-inflammatory cytokines such as TNF-α [86] and activate NKT cells to produce IFN-γ by presenting glycolipid antigens [81]. To test the role of the $A_{2A}R$ expressed on DCs in kidney IRI, CD11c-Cre $A_{2A}R^{fl/fl}$ mice (which are deficient in $A_{2A}R$ expression in CD11c+ DCs) were generated and exposed to a mild ischemic insult. These CD11c-Cre $A_{2A}R^{fl/fl}$ mice developed severe renal dysfunction compared to CD11c-Cre control mice exposed to the same length of ischemia [38]. Furthermore, whereas $A_{2A}R$ agonist treatment protected the CD11c-Cre control mice from IRI, no protection was observed in the CD11c-Cre $A_{2A}R^{fl/fl}$ mice.

The powerful ability of $A_{2A}R$ activation on DCs to protect the kidney was further demonstrated in experiments where bone marrow-derived DCs (BMDCs) were cultured in the presence of an $A_{2A}R$ agonist or vehicle and loaded with the NKT cell-specific glycolipid antigen α-galactosylceramide (α-GalCer). Adoptive transfer of the α-GalCer-loaded $A_{2A}R$ agonist-treated BMDCs prior to, and up to 6 hours after a severe ischemic insult, conferred significant protection from renal dysfunction and acute tubular necrosis [38].

On the other hand, α-GalCer-loaded vehicle-treated BMDCs either had no effect on renal injury (after a severe ischemic insult) or exacerbated renal injury (after a mild ischemic insult) [38]. The mechanism of action for the α-GalCer-loaded $A_{2A}R$ agonist-treated BMDCs was found to involve the stimulation of IL-10 production by cells in the mice receiving adoptive transfer [38]. In summary, intrinsic DCs are a target of endogenous adenosine via $A_{2A}R$ as well as specific $A_{2A}R$ agonists administered to protect the kidney from IRI. Furthermore, $A_{2A}R$ stimulation on BMDCs *ex vivo* can endow these cells with the ability to suppress kidney IRI in recipient mice upon adoptive transfer.

Studies in other models of inflammation reported that $A_{2A}R$ activation inhibits neutrophil adhesion to the endothelium or other cell types [87, 88], and a damaging role for neutrophils in kidney IRI has been reported [89]. $A_{2A}R$ agonist treatment significantly reduced neutrophil transmigration from the blood into the extravascular space in the kidney after ischemia [90]. This finding suggests that $A_{2A}R$ agonists may also directly act on neutrophils during reperfusion to inhibit neutrophil extravasation into kidney tissue and reduce further damage.

Additional studies have investigated the role of $A_{2A}Rs$ on monocytes/macrophages in kidney IRI [91]. Day et al. used liposomal clodronate to deplete monocytes/macrophages from WT mice that were then reconstituted with an $A_{2A}R$-deficient macrophage cell line (RAW 264.7 cells treated with $A_{2A}R$ siRNA) [91]. In these mice the $A_{2A}R$ agonist ATL146e markedly protected against kidney IRI. In the converse experiment, where $A_{2A}R$ KO mice were depleted of monocytes/macrophages and reconstituted with WT macrophages (RAW 264.7 not treated with $A_{2A}R$ siRNA), the $A_{2A}R$ agonist lost all protective properties [91]. These results suggest that monocytes/macrophages are not a direct target of $A_{2A}R$ agonists in kidney IRI.

Experimental evidence thus far suggests that the major protective effects of $A_{2A}R$ agonists in kidney IRI are mediated through modulation of immune cell functions. This has been studied by generation of bone marrow chimeric mice and other mouse models where the $A_{2A}R$ is removed from one type of immune cell but not others. Important kidney-protective contributions are made by CD4+ T cells (Tregs and NKT cells), DCs and neutrophils expressing the $A_{2A}R$. On the other hand, the role of $A_{2A}Rs$ on monocytes/macrophages appears to be minimal. Overall, $A_{2A}R$ agonists are a promising potential therapeutic strategy for kidney IRI.

1.5. Role of $A_{2A}R$ in Cardiac IRI

IRI to the heart differs from that to the lung and kidney due not only to differences in tissue architecture but also due to the higher metabolic demands of cardiomyocytes. Because of this, even brief periods of ischemia (as little as 15 min) can lead to appreciable sarcolemmal rupture and cardiomyocyte necrosis, followed sequentially by neutrophil infiltration, macrophage debridement, myofibroblast proliferation, neovascularization, collagen deposition and maturation of the infarct scar [92]. The cardioprotective effects of adenosine itself have long been appreciated [93], and many genetic and pharmacologic studies conducted over the years have implicated each of the four adenosine receptors in cardiac IRI [94]. Much of the early work focused on adenosine's ability to trigger a preconditioned state through the A_1R [95]. Nevertheless, studies soon followed establishing cardioprotective roles for the $A_{2A}R$ [96] and A_3R [97], and more recently a cardioprotective role for the $A_{2B}R$ has been suggested [98]. Thus each of the four adenosine receptors have been implicated in cardiac IRI.

As mentioned above, the cardioprotective effects of adenosine were originally understood in terms of adenosine's ability to trigger a preconditioned state through the A_1R [95]. But while ischemic preconditioning is primarily associated with a direct effect on cardiomyocytes, the protective effects of adenosine during reperfusion were associated with a decrease in the influx of neutrophils into the ischemic zone and with reduced damage to the endothelium [93]. Additional studies conducted with selective adenosine receptor agonists showed that this form of cardioprotection could largely be attributed to $A_{2A}R$ stimulation [25, 99]. Since that time, the administration of selective $A_{2A}R$ agonists at (or shortly before) reperfusion has been shown to reduce infarct size in a wide variety of animal models including pigs [100], dogs [101, 102], rabbits [103], rats [104] and mice [105]. Importantly, the cardioprotection provided by selective $A_{2A}R$ agonists is lost in $A_{2A}R$ knockout mice (105), as well as in lymphocyte-deficient Rag-1 knockout mice reconstituted with CD4+ T cells from $A_{2A}R$ knockout mice, which indicates that $A_{2A}Rs$ on CD4+ T cells are primarily responsible for the cardioprotective effects of $A_{2A}R$ agonists [34].

One of the longstanding limitations of ischemic preconditioning is that it cannot be translated directly to clinical practice given that it is essentially impossible to predict when a patient might suffer an ischemic event. In contrast, the efficacy of $A_{2A}R$ agonists administered minutes before reperfusion makes them directly translatable to the current practice of direct percutaneous intervention (PCI) for acute myocardial infarction. Thus the timing of $A_{2A}R$ administration for infarct size reduction is similar to that of ischemic postconditioning [106]. As in ischemic preconditioning, postconditioning also involves a series of brief ischemic episodes interrupted by reperfusion, however postconditioning entails a series of ischemic episodes imposed at the end of the prolonged coronary occlusion (rather than at the beginning as in preconditioning). Whether ischemic postconditioning functions primarily to protect cardiomyocytes from calcium overload and opening of the mitochondrial permeability transition pore (mPTP), or by suppressing inflammatory responses during reperfusion will depend to some extent upon the duration of the ischemic event. Nevertheless, postconditioning is known to both inhibit neutrophil infiltration and to preserve endothelial function [107]. From the perspective of this review, it is also noteworthy that the cardioprotective effects of postconditioning are greatly reduced in $A_{2A}R$ knockout mice [108]. Taken together with the finding that the cardioprotective effects of $A_{2A}R$ agonists are primarily mediated through CD4+ T cells [34], this would indicate that both ischemic postconditioning and $A_{2A}R$ agonists function primarily by inhibiting the additional injury imposed by innate inflammatory responses during reperfusion of the heart.

1.6. Role of $A_{2A}R$ in Liver IRI

Liver, being an organ with high energy requirements, is highly dependent on oxygen supply and susceptible to hypoxic or ischemic conditions. Hepatic IRI occurs substantially during liver resection or transplantation and remains a major cause of liver dysfunction following liver surgery for which the only effective treatment is re-transplantation. The major consequences of liver IRI are hepatocyte and sinusoidal endothelial cell necrosis, apoptosis or a combination of both [109]. This injury has become a bottleneck which has restricted the use of marginal liver donors and the development of extensive liver resection. Mechanisms involved in hepatic IRI include anaerobic metabolism, oxidative stress, intracellular calcium overload, activation of Kupffer cells and neutrophils, and production of proinflammatory cytokines and chemokines. More importantly, an effective prevention or treatment method is still lacking.

Adenosine has been shown to prevent liver IRI via activation of $A_{2A}R$. Ben-Ari et al. demonstrated that administration of CGS21680, an $A_{2A}R$ agonist, before induction of ischemia attenuated postischemic apoptotic hepatic injury, possibly mediated through caspase-3 activity [110]. These results showed that at 1 minute of reperfusion there was significant reduction in liver enzyme levels and caspase-3 activity in animals pretreated with CGS21680. Also, reduction in postischemic apoptotic hepatic injury in the CGS21680-treated group was confirmed morphologically by the demonstration of significantly fewer apoptotic hepatocyte cells. Harada et al. demonstrated that adenosine and selective $A_{2A}R$ agonists reduced IRI of rat liver largely by inhibiting leukocyte activation via inhibiting the synthesis of inflammatory mediators and inhibiting neutrophil degranulation directly [111].

The mechanism of concanavalin A (ConA)-induced acute liver injury involves CD4+ T cell activation accompanied by the secretion of proinflammatory cytokines such as TNF-α, IFN-γ and IL-6, resulting in liver cell damage, apoptosis, necrosis and ultimately liver death [112-114]. Both elevated serum cytokine levels and liver injury have been shown to be inhibited by the $A_{2A}R$ agonist ATL146e [91, 112, 113], an effect not observed in $A_{2A}R$ KO mice, which displayed enhanced ConA-induced liver inflammation and necrosis [91, 115]. The ability of $A_{2A}R$ agonists to reduce inflammation and afford tissue protection in hepatic IRI via inhibition of proinflammatory cytokine production has been demonstrated in numerous studies [33, 91, 113, 116, 117]. For instance, ATL146e administered at the time of reperfusion suppressed the production of numerous cytokines in WT mice but not in $A_{2A}R$ KO mice, likely contributing to tissue protection since this effect was blocked by the selective $A_{2A}R$ antagonist ZM241385 [91, 117]. Similarly, Hasko et al. demonstrated that $A_{2A}R$ activation inhibited the secretion of IL-12 and TNF-α [118]. Interestingly, expression of $A_{2A}R$ increased following small-for-size liver transplantation, suppressing NF-κB activation and decreasing the expression of TNF-α, CXCL2 and ICAM-1 [119].

Most of these studies clearly demonstrated that protection afforded by $A_{2A}R$ agonists was largely attributable to the activation of $A_{2A}Rs$ on CD4+ T cells and to the inhibition of T cell receptor (TCR)-mediated IFN-γ production. However, the rapidity of IRI led Lappas et al. to speculate whether activation of the NKT cell subset, by a CD1d-restricted mechanism, played an important role in initiating the inflammatory cascade and was the target of $A_{2A}R$ agonists [33]. NKT cells recognize lipid antigens derived from pathogens or possibly damaged tissue that are presented by the major histocompatibility complex class I-related molecule CD1d, by antigen presenting cells. Lappas et al. demonstrated that NKT cells were stimulated to produce IFN-γ within two hours of reperfusion. The use of antibodies to deplete NK1.1+ cells (NK and NKT cells) or to block CD1d-restricted antigen presentation to NKT cells replicated, but was not additive to, the protection afforded by ATL146e. Also, the adoptive transfer of NKT cells purified from WT or $A_{2A}R$ KO mice but not IFN-γ KO mice, restored IRI to the WT level in Rag-1 KO mice (which lack T and B cells). Rag-1 KO mice that were reconstituted with $A_{2A}R$ KO NKT cells were not protected from IRI by ATL146e. Also, ATL146e was capable of potently inhibiting IFN-γ production in NKT cells in response to anti-CD3 antibodies or the CD1d-restricted glycolipid antigen, α-GalCer. Therefore, it appears that NKT cell activation initiates the recruitment and transactivation of other immune cells such as macrophages and neutrophils that propagates tissue inflammation and injury after IR.

It is now known that all four types of adenosine receptors are present on mouse NKT cells, and while adenosine can suppress IFN-γ production by NKT cells, $A_{2A}R$ activation is required for the secretion of IL-4 and IL-10 by these cells [120]. This is supported by the observation that $A_{2A}R$ KO mice exhibited largely decreased levels of IL-4, IL-10 and TGF-β concomitantly with an increase of IFN-γ upon α-GalCer administration *in vivo* [120]. Thus, while $A_{2A}R$ activation inhibits other lymphocyte populations, it is required for the secretion of IL-4 and IL-10 by NKT cells. Interestingly, Alchera et al. have demonstrated that activation of $A_{2A}Rs$ induces a late preconditioning in liver cells through the activation of hypoxia inducible factor-1 (HIF-1). HIF-1-induced expression of carbonic anhydrase IX, a HIF-1-regulated transmembrane enzyme that catalyzes bicarbonate production, increased hepatocyte tolerance to ischemia by amelioration of intracellular acidosis and maintenance of intracellular Na+ homeostasis [121]. Chouker et al. have also demonstrated that a total body hypoxia-triggered pathway provided protection in acute hepatitis, and that hypoxia (upstream) and $A_{2A}R$ (downstream) functioned in the same immunosuppressive and liver tissue-protecting pathway [115]. It was also shown that *in vivo* hypoxic preconditioning was protective against 'warm' liver IRI through the $A_{2B}R$ [122].

1.7. Role of $A_{2A}R$ in Pancreas IRI

Little is known about the immunosuppressive role of $A_{2A}R$ activation in autoimmune type 1 diabetes or in promoting islet graft survival and reducing islet loss post-transplantation. Sakowicz-Burkiewicz et al. used specific adenosine receptor agonists and antagonists to demonstrate that adenosine-induced suppression of diabetic T cell proliferation was mediated by the $A_{2A}R$ and not the $A_{2B}R$ [123]. Decreased expression of adenosine kinase was shown to suppress the proliferation of diabetic T lymphocytes, leading to adenosine out-flow from the cell. The extracellular adenosine then stimulated $A_{2A}R$ and induced cAMP production, leading to the activation of PKA and the suppression of T cell proliferation. Administration of CGS21680 in a rat model of severe, taurocholate-induced experimental acute pancreatitis (EAP) was also shown to favorably decrease inflammatory cell infiltration, hemorrhagic changes, necrosis and vacuolization of acinar cells, without an evident effect on edema of the interstitial tissue, suggesting that modulation by $A_{2A}R$ agonist could help alleviate severe EAP [124]. Similarly, in diabetic rats, chronic $A_{2A}R$ activation was found to ameliorate histological and functional changes in kidneys induced by diabetes and to reduce inflammation associated with diabetic nephropathy [125]. However, it is interesting to note that while $A_{2A}R$ agonists are potential candidates for the treatment of type 1 diabetes through immunomodulation, $A_{2B}Rs$ also play a significant role in these events [126].

Zarek et al. demonstrated that $A_{2A}R$ stimulation not only inhibited the generation of adaptive Th1 and Th17 effector cells and promoted long-term T-cell anergy, but also promoted the induction of Treg cells, an observation with immense clinical significance given the involvement of these subsets in transplant rejection and tolerance, respectively [127]. In their investigation of the effect of adenosine therapy in islet transplantation, Nitta et al. demonstrated that a single dose of adenosine delivered at the time of islet transplantation suppressed NKT cell-mediated IFN-γ production by neutrophils in the liver of streptozotocin-induced diabetic recipient mice, thus preventing early loss of transplanted syngeneic and allogeneic islets and enabling transplantation using a single donor [128]. This observation, combined with the observation of Lappas et al. that $A_{2A}R$ agonists can suppress hepatic IRI by inhibiting NKT cell activation [33], clearly suggests an important involvement of NKT cells in the early loss of transplanted islets. However, whether the mechanism of inhibition lies upstream of NKT cell activation, with an involvement of DCs and macrophages, is yet to be elucidated. In an attempt to further determine the extent and efficiency of improvement in islet transplantation by adenosine administration, an intra-peritoneal glucose tolerance test performed 60 days post-transplantation showed that the glucose tolerance of diabetic mice receiving 200 islets from a single donor and treated once with adenosine was superior in comparison with diabetic mice receiving 400 islets from two donors, indicating that adenosine facilitated a greater than 2-fold improvement in the efficiency of islet transplantation [128]. The beneficial effect of adenosine therapy was also observed in islet allotransplantation when treatment included an immunosuppressive agent such as anti-CD4 antibody. This study supports an earlier observation from the same group that survival of islet allograft in the liver was prolonged without any immunosuppression in NKT cell- and CD1d-deficient mice, indicating an essential role for NKT cells in alloimmune rejection of islets [129, 130]. Interestingly, they also showed that successful islet transplantation from one donor to two recipients was feasible by targeting pro-inflammatory cytokines in mice: restoration of normoglycemia was achieved in WT diabetic mice receiving 100 syngeneic islet grafts by administration of anti-TNF-α and anti-IL-1β antibodies in conjunction with anti-IFN-γ antibody. Transplantation of 200 syngeneic islets from a single donor pancreas into IFN-γ KO mice produced similar results. Normoglycemia was also attained by the combined use of antibodies in mice receiving islet allografts when rejection was prevented with anti-CD4 antibody treatment. However, the role of $A_{2A}R$ agonist-mediated inhibition of the cytotoxic activity and cytokine production by activated NKT cells in islet transplantation has yet to be elucidated. A recent study by Oskouie et al. demonstrated that upregulation and subsequent activation of adenosine deaminase in NOD DCs triggers autoimmune T cell activation and diabetes in NOD mice [131]. DCs deficient in adenosine deaminase expression have a greatly reduced capacity to trigger type 1 diabetes. Also interesting is the observation that expression of human CD39 on mouse islets is sufficient to protect against chemically-induced diabetes [132]. Furthermore, the level of hCD39 expression correlates inversely with severity of hyperglycemia.

Studies by Chhabra et al., conducted in both the murine sub-kidney capsule as well as the clinically-relevant hepatic intra-portal islet transplantation models, indicate that infusion of $A_{2A}R$ agonists ATL146e or ATL313 in diabetic recipient mice during the peritransplant period results in reduction of 1) inflammation-mediated early loss of a minimal dose islet graft and 2) the time taken to achieve sustained normoglycemia [133, 134]. No significant improvement in time taken to achieve normoglycemia was observed between diabetic $A_{2A}R$ KO recipients of congenic WT islets with or without administration of ATL146e therapy. In histological studies, $A_{2A}R$ agonist treatment demonstrated marked reductions in neutrophil and lymphocyte infiltration, cellular necrosis and fibrosis, as well as in the size and number of lymphoid nodules. *In vitro* studies indicated that islets incubated with $A_{2A}R$ agonists remained viable and functional. Co-culture of islets with leukocytes followed by glucose-stimulated islet secretion challenge demonstrated an inhibition of islet stimulation index that was subsequently restored upon addition of ATL146e. Furthermore, this beneficial effect was blocked by addition of an $A_{2A}R$ selective antagonist SCH58261, indicating that ATL146e/$A_{2A}R$ binding on inflammatory cells inhibited inflammation-mediated islet cell damage and dysfunction. This was supported by the observed failure of ATL146e to preserve the insulin secretion function of islets co-cultured with leukocytes from $A_{2A}R$ KO mice. Therefore, these results indicate that in islet transplant protocols designed to minimize systemic immunosuppression, including a short course of $A_{2A}R$ agonist treatment as part of induction therapy, could reduce early peritransplant islet loss, thereby improving islet graft outcomes. Additionally, by reducing the number of islets required for successful clinical islet transplantation, a greater efficiency in the use of human islets for clinical transplantation would increase the availability of this procedure to a greater number of potential recipients. In conclusion, these studies suggest that $A_{2A}R$ agonist therapy reduces inflammation-mediated islet cell loss following transplantation, thereby allowing for an accelerated rate of cure with marginal donor islets. Including a short course of $A_{2A}R$ agonist administration as part of the induction therapy in immunosuppressive regimens would increase both the efficacy of the procedure in terms of improved long term islet transplantation outcomes as well as the availability of the procedure to a larger number of diabetic recipients.

1.8. Clinical Experience with $A_{2A}R$ Agonists

Although $A_{2A}R$ agonists have not yet been tested clinically in organ transplantation, over recent years a number of $A_{2A}R$ agonists have been evaluated in clinical trials as anti-inflammatory agents for a variety of indications (Table 1). For the most part, the results from these anti-inflammatory trials have been disappointing because the compounds were not shown to be clinically efficacious [135, 136]. MRE-0094 (King Pharmaceuticals, Inc.), GW-328267X (GlaxoSmithKline) and UK 432,097 (Pfizer) all failed to meet their primary study endpoints. Unfortunately, it is difficult to determine the cause for the failure of these studies because results from failed clinical trials are often not reported. One possibility for the failures could be that the compounds were not as selective as they were believed to be based on *in vitro* data. Binding studies conducted using non-coupled $A_{2A}R$ cell assays do not reflect the high-affinity state of the receptor and thus adversely affect the estimation of the relative subtype selectivity [137]. Another possibility is that the doses administered produced unwanted events such as hypotension or tachycardia, well known effects of $A_{2A}R$ stimulation at higher doses.

In addition to potential use as treatments for inflammation, three $A_{2A}R$ agonists, regadenoson (CVT3146, Lexiscan™), binodenoson (MRE-0470) and apadenoson (ATL146e, Stedivaze™), have undergone human testing as pharmacologic stress agents for myocardial perfusion imaging studies in patients that are unable to exercise. Of these agents, regadenoson became the first $A_{2A}R$ agonist to be approved by the FDA in July, 2008, and apadenoson is currently in Phase 3 testing for this indication. Binodenoson completed Phase 3 in 2008 and demonstrated efficacy and improved tolerability versus adenosine but was ultimately not approved by the FDA due to problems with the trial design. Although approved by the FDA for use in imaging studies, regadenoson has also been evaluated off-label as an anti-inflammatory agent in sickle cell disease patients during painful vaso-occlusive crises (pVOC) [55]. In a Phase 1 dose-seeking and safety study that was completed in March, 2013, the investigators demonstrated that pVOC causes activation of NKT cells and that a low-dose infusion of regadenoson (1.44 µg/kg/hr x 24 hrs) decreases NKT cell activation without toxicity. Importantly, this dose had no appreciable effect on heart rate or blood pressure, and no cardiovascular toxicities were identified. Based on these findings, two Phase 2 clinical trials are currently recruiting patients with sickle cell disease for further testing.

Three other $A_{2A}R$ agonists have completed early stage clinical trials. OPA-6566 (Acucela/Otsuka Pharmaceutical) and ATL313 (Clinical Data/Santen Pharmaceutical) were investigated for use as a topical ophthalmic solution in patients with open-angle glaucoma or ocular hypertension. The mechanism of action is by $A_{2A}R$ stimulation of aqueous humor outflow via the trabecular meshwork. OPA-6566 remains in the Acucela pipeline but ATL313 development has been discontinued for this indication. Biovitrum reported positive Phase 2 clinical trial results using a pH-sensitive $A_{2A}R$ agonist, BVT.115959, as an oral treatment for diabetic neuropathic pain. BVT.115959 is reported to have a higher affinity for the $A_{2A}R$ at pH 7.0 than at 7.4, giving it some measure of selectivity for inflamed tissues where the local pH is reduced. The mechanism of action for this compound as a treatment for diabetic neuropathic pain is both vasodilatory and anti-inflammatory at the sites of the distal nerve endings.

Table 1. $A_{2A}R$ agonists used in anti-inflammatory studies

Pre-Clinical Compounds		Company/ Institution	References		
CGS-21680		Catalog compound	(52, 53, 110, 111, 124, 138, 139)		
ATL-146e, ATL-313, ATL-370, ATL-1222, ATL-1223		Adenosine Therapeutics, LLC	(35, 45, 80, 91, 101, 140, 141)		
Clinical Compounds	Indication	Company/ Institution	Clinical Develop-ment Stage	Status	ClinicalTrials.gov Identifier
MRE-0094 (sonedenoson)	Wound Healing (Diabetic foot ulcers)	King Pharmaceuticals, Inc. and Pfizer	Phase 2	Completed	NCT00312364, NCT00318214
PDRN	Wound Healing (Diabetic foot ulcers)	University of Messina	Phase 4	Completed	NCT00638872
GW-328267X	Allergic rhinitis, asthma	GlaxoSmithKline	Phase 2a	Completed	NCT01640990
UK432,097	COPD	Pfizer	Phase 2b	Terminated	NCT00430300
CVT-3146 (regadenoson)	Sickle cell crisis	Dana Farber Cancer Institute	Phase 1	Completed	NCT01085201
CVT-3146 (regadenoson)	Sickle cell crisis	Dana Farber Cancer Institute, Blood Center of Wisconsin	Phase 2	Recruiting	NCT01788631, NCT01566890

Conclusion

The activation of $A_{2A}R$ signaling pathways plays a crucial role in preventing acute tissue injury after IR via potent anti-inflammatory and immunosuppressive actions. In particular, the role of hypoxia and hypoxia-associated accumulation of extracellular adenosine characterizes the most fundamental process of tissue damage, which can be attenuated through the inhibition of activated immune cells by $A_{2A}R$ agonists. The pivotal role of hypoxia-induced extracellular adenosine and $A_{2A}R$ signaling to regulate innate and adaptive immune systems during IRI underlines the importance of this signaling pathway in organ transplantation. The multi-faceted role of $A_{2A}R$

activation involves signal transduction pathways on multiple cellular targets for their effector responses. In particular, $A_{2A}R$ inhibits CD4+ NKT cell activation thereby limiting cytokine output leading to inflammation. The immunosuppressive phenotype of M2 macrophages and tolerogenic DCs can also be modulated by $A_{2A}R$-mediated effector mechanisms. The effect of $A_{2A}R$ activation on proliferation of anti-inflammatory cell types such as Tregs and myeloid-derived suppressor cells (MDSCs) as well as the release of cytokines and immune modulatory factors such as VEGF, IL-10, TGF-β and IL-6 can also lead to enhanced immunosurveillance and enhanced graft survival after organ transplantation. Thus targeting of $A_{2A}Rs$ by pharmacologic agonists represents a promising therapeutic strategy for the prevention and treatment of acute graft injury after transplantation. The use of $A_{2A}R$ agonists with other anti-inflammatory drugs that target alternative cellular pathways may have synergistic effects, representing a novel and effective means for the therapeutic management of IRI after organ transplantation.

References

[1] Anderson CM, Xiong W, Young JD, Cass CE, Parkinson FE. Demonstration of the existence of mRNAs encoding N1/cif and N2/cit sodium/nucleoside cotransporters in rat brain. *Brain Res. Mol. Brain Res.*, 42:358-361, 1996.

[2] Huang QQ, Yao SY, Ritzel MW, Paterson AR, Cass CE, Young JD. Cloning and functional expression of a complementary DNA encoding a mammalian nucleoside transport protein. *J. Biol. Chem.*, 269:17757-17760, 1994.

[3] Williams TC, Jarvis SM. Multiple sodium-dependent nucleoside transport systems in bovine renal brush-border membrane vesicles. *Biochem. J.*, 274 (Pt 1):27-33, 1991.

[4] Arch JR, Newsholme EA. Activities and some properties of 5'-nucleotidase, adenosine kinase and adenosine deaminase in tissues from vertebrates and invertebrates in relation to the control of the concentration and the physiological role of adenosine. *Biochem. J.*, 174:965-977, 1978.

[5] Lloyd HG, Fredholm BB. Involvement of adenosine deaminase and adenosine kinase in regulating extracellular adenosine concentration in rat hippocampal slices. *Neurochem. Int.*, 26:387-395, 1995.

[6] Latini S, Bordoni F, Corradetti R, Pepeu G, Pedata F. Effect of A2A adenosine receptor stimulation and antagonism on synaptic depression induced by in vitro ischaemia in rat hippocampal slices. *Br. J. Pharmacol.,* 128:1035-1044, 1999.

[7] Latini S, Bordoni F, Pedata F, Corradetti R. Extracellular adenosine concentrations during in vitro ischaemia in rat hippocampal slices. *Br. J. Pharmacol.,* 127:729-739, 1999.

[8] Rudolphi KA, Schubert P, Parkinson FE, Fredholm BB. Adenosine and brain ischemia. *Cerebrovasc. Brain Metab. Rev.*, 4:346-369, 1992.

[9] Rudolphi KA, Schubert P, Parkinson FE, Fredholm BB. Neuroprotective role of adenosine in cerebral ischaemia. *Trends Pharmacol. Sci.*, 13:439-445, 1992.

[10] Vizi E, Huszar E, Csoma Z, Boszormenyi-Nagy G, Barat E, Horvath I, Herjavecz I, Kollai M. Plasma adenosine concentration increases during exercise: a possible contributing factor in exercise-induced bronchoconstriction in asthma. *J. Allergy Clin. Immunol.,* 109:446-448, 2002.

[11] Decking UK, Schlieper G, Kroll K, Schrader J. Hypoxia-induced inhibition of adenosine kinase potentiates cardiac adenosine release. *Circ. Res.,* 81:154-164, 1997.

[12] Fredholm BB, AP IJ, Jacobson KA, Klotz KN, Linden J. International Union of Pharmacology. XXV. Nomenclature and classification of adenosine receptors. *Pharmacol. Rev.,* 53:527-552, 2001.

[13] Hasko G, Linden J, Cronstein B, Pacher P. Adenosine receptors: therapeutic aspects for inflammatory and immune diseases. *Nat. Rev. Drug Discov.,* 7:759-770, 2008.

[14] Fredholm BB, Irenius E, Kull B, Schulte G. Comparison of the potency of adenosine as an agonist at human adenosine receptors expressed in Chinese hamster ovary cells. *Biochem. Pharmacol.,* 61:443-448, 2001.

[15] Fredholm BB. Adenosine, an endogenous distress signal, modulates tissue damage and repair. *Cell Death Differ.,* 14:1315-1323, 2007.

[16] Hasko G, Cronstein BN. Adenosine: an endogenous regulator of innate immunity. *Trends Immunol.,* 25:33-39, 2004.

[17] Newby AC, Worku Y, Holmquist CA. Adenosine formation. Evidence for a direct biochemical link with energy metabolism. *Adv. Myocardiol.,* 6:273-284, 1985.

[18] Murphree LJ, Sullivan GW, Marshall MA, Linden J. Lipopolysaccharide rapidly modifies adenosine receptor transcripts in murine and human

macrophages: role of NF-kappaB in A(2A) adenosine receptor induction. *Biochem. J.,* 391:575-580, 2005.

[19] Hasko G, Pacher P. A2A receptors in inflammation and injury: lessons learned from transgenic animals. *J. Leukoc. Biol.,* 83:447-455, 2008.

[20] Cronstein BN, Kramer SB, Rosenstein ED, Korchak HM, Weissmann G, Hirschhorn R. Occupancy of adenosine receptors raises cyclic AMP alone and in synergy with occupancy of chemoattractant receptors and inhibits membrane depolarization. *Biochem. J.,* 252:709-715, 1988.

[21] Thiel M, Bardenheuer H, Poch G, Madel C, Peter K. Pentoxifylline does not act via adenosine receptors in the inhibition of the superoxide anion production of human polymorphonuclear leukocytes. *Biochem. Biophys. Res. Commun.,* 180:53-58, 1991.

[22] Sullivan GW, Rieger JM, Scheld WM, Macdonald TL, Linden J. Cyclic AMP-dependent inhibition of human neutrophil oxidative activity by substituted 2-propynylcyclohexyl adenosine A(2A) receptor agonists. *Br. J. Pharmacol.,* 132:1017-1026, 2001.

[23] Milne GR, Palmer TM. Anti-inflammatory and immunosuppressive effects of the A2A adenosine receptor. *Scientific World Journal,* 11:320-339, 2011.

[24] Kreckler LM, Wan TC, Ge ZD, Auchampach JA. Adenosine inhibits tumor necrosis factor-alpha release from mouse peritoneal macrophages via A2A and A2B but not the A3 adenosine receptor. *J. Pharmacol. Exp. Ther.,* 317:172-180, 2006.

[25] Nakamura M, Zhao ZQ, Clark KL, Velez DV, Guyton RA, Vinten-Johansen J. A novel adenosine analog, AMP579, inhibits neutrophil activation, adherence and neutrophil-mediated injury to coronary vascular endothelium. *Eur. J. Pharmacol.,* 397:197-205, 2000.

[26] Fredholm BB, Zhang Y, van der Ploeg I. Adenosine A2A receptors mediate the inhibitory effect of adenosine on formyl-Met-Leu-Phe-stimulated respiratory burst in neutrophil leucocytes. *Naunyn. Schmiedebergs Arch. Pharmacol.,* 354:262-267, 1996.

[27] Ledent C, Vaugeois JM, Schiffmann SN, Pedrazzini T, El Yacoubi M, Vanderhaeghen JJ, Costentin J, Heath JK, Vassart G, Parmentier M. Aggressiveness, hypoalgesia and high blood pressure in mice lacking the adenosine A2a receptor. *Nature,* 388:674-678, 1997.

[28] Sexl V, Mancusi G, Holler C, Gloria-Maercker E, Schutz W, Freissmuth M. Stimulation of the mitogen-activated protein kinase via the A2A-adenosine receptor in primary human endothelial cells. *J. Biol. Chem.,* 272:5792-5799, 1997.

[29] Nadeem A, Fan M, Ansari HR, Ledent C, Jamal Mustafa S. Enhanced airway reactivity and inflammation in A2A adenosine receptor-deficient allergic mice. *Am. J. Physiol. Lung Cell Mol. Physiol.,* 292:L1335-1344, 2007.

[30] Lukashev D, Ohta A, Sitkovsky M. Targeting hypoxia--A(2A) adenosine receptor-mediated mechanisms of tissue protection. *Drug Discov. Today,* 9:403-409, 2004.

[31] Sands WA, Woolson HD, Milne GR, Rutherford C, Palmer TM. Exchange protein activated by cyclic AMP (Epac)-mediated induction of suppressor of cytokine signaling 3 (SOCS-3) in vascular endothelial cells. *Mol. Cell Biol.,* 26:6333-6346, 2006.

[32] Day YJ, Huang L, McDuffie MJ, Rosin DL, Ye H, Chen JF, Schwarzschild MA, Fink JS, Linden J, Okusa MD. Renal protection from ischemia mediated by A2A adenosine receptors on bone marrow-derived cells. *J. Clin. Invest.,* 112:883-891, 2003.

[33] Lappas CM, Day YJ, Marshall MA, Engelhard VH, Linden J. Adenosine A2A receptor activation reduces hepatic ischemia reperfusion injury by inhibiting CD1d-dependent NKT cell activation. *J. Exp. Med.,* 203:2639-2648, 2006.

[34] Yang Z, Day YJ, Toufektsian MC, Xu Y, Ramos SI, Marshall MA, French BA, Linden J. Myocardial infarct-sparing effect of adenosine A2A receptor activation is due to its action on CD4+ T lymphocytes. *Circulation,* 114:2056-2064, 2006.

[35] Sharma AK, Laubach VE, Ramos SI, Zhao Y, Stukenborg G, Linden J, Kron IL, Yang Z. Adenosine A2A receptor activation on CD4+ T lymphocytes and neutrophils attenuates lung ischemia-reperfusion injury. *J. Thorac. Cardiovasc. Surg.,* 139:474-482, 2010.

[36] Sharma AK, LaPar DJ, Zhao Y, Li L, Lau CL, Kron IL, Iwakura Y, Okusa MD, Laubach VE. Natural killer T cell-derived IL-17 mediates lung ischemia-reperfusion injury. *Am. J. Respir. Crit. Care Med.,* 183:1539-1549, 2011.

[37] Khoa ND, Montesinos MC, Reiss AB, Delano D, Awadallah N, Cronstein BN. Inflammatory cytokines regulate function and expression of adenosine A(2A) receptors in human monocytic THP-1 cells. *J. Immunol.,* 167:4026-4032, 2001.

[38] Li L, Huang L, Ye H, Song SP, Bajwa A, Lee SJ, Moser EK, Jaworska K, Kinsey GR, Day YJ, Linden J, Lobo PI, Rosin DL, Okusa MD. Dendritic cells tolerized with adenosine A(2)AR agonist attenuate acute kidney injury. *J. Clin. Invest.,* 122:3931-3942, 2012.

[39] Christie JD, Edwards LB, Kucheryavaya AY, Aurora P, Dobbels F, Kirk R, Rahmel AO, Stehlik J, Hertz MI. The Registry of the International Society for Heart and Lung Transplantation: twenty-seventh official adult lung and heart-lung transplant report--2010. *J. Heart Lung Transplant.,* 29:1104-1118, 2010.

[40] McGregor CG, Daly RC, Peters SG, Midthun DE, Scott JP, Allen MS, Tazelaar HD, Keating MR, Walker RC, McDougall JC. Evolving strategies in lung transplantation for emphysema. *Ann. Thorac. Surg.,* 57:1513-1520, 1994.

[41] Aeba R, Griffith BP, Kormos RL, Armitage JM, Gasior TA, Fuhrman CR, Yousem SA, Hardesty RL. Effect of cardiopulmonary bypass on early graft dysfunction in clinical lung transplantation. *Ann. Thorac. Surg.,* 57:715-722, 1994.

[42] Zenati M, Yousem SA, Dowling RD, Stein KL, Griffith BP. Primary graft failure following pulmonary transplantation. *Transplantation,* 50:165-167, 1990.

[43] Fiser SM, Tribble CG, Long SM, Kaza AK, Kern JA, Jones DR, Robbins MK, Kron IL. Ischemia-reperfusion injury after lung transplantation increases risk of late bronchiolitis obliterans syndrome. *Ann. Thorac. Surg.,* 73:1041-1047, 2002.

[44] Lau CL, Zhao Y, Kron IL, Stoler MH, Laubach VE, Ailawadi G, Linden J. The role of adenosine A2A receptor signaling in bronchiolitis obliterans. *Ann. Thorac. Surg.,* 88:1071-1078, 2009.

[45] LaPar DJ, Laubach VE, Emaminia A, Crosby IK, Hajzus VA, Sharma AK, Sumner HM, Webb DV, Lau CL, Kron IL. Pretreatment strategy with adenosine A2A receptor agonist attenuates reperfusion injury in a preclinical porcine lung transplantation model. *J. Thorac. Cardiovasc. Surg.,* 142:887-894, 2011.

[46] Sharma AK, Linden J, Kron IL, Laubach VE. Protection from pulmonary ischemia-reperfusion injury by adenosine A2A receptor activation. *Respir. Res.,* 10:58, 2009.

[47] Ross SD, Tribble CG, Linden J, Gangemi JJ, Lanpher BC, Wang AY, Kron IL. Selective adenosine-A2A activation reduces lung reperfusion injury following transplantation. *J. Heart Lung Transplant,* 18:994-1002, 1999.

[48] Gazoni LM, Laubach VE, Mulloy DP, Bellizzi A, Unger EB, Linden J, Ellman PI, Lisle TC, Kron IL. Additive protection against lung ischemia-reperfusion injury by adenosine A2A receptor activation

before procurement and during reperfusion. *J. Thorac. Cardiovasc. Surg.,* 135:156-165, 2008.

[49] Rivo J, Zeira E, Galun E, Einav S, Linden J, Matot I. Attenuation of reperfusion lung injury and apoptosis by A2A adenosine receptor activation is associated with modulation of Bcl-2 and Bax expression and activation of extracellular signal-regulated kinases. *Shock,* 27:266-273, 2007.

[50] Reece TB, Ellman PI, Maxey TS, Crosby IK, Warren PS, Chong TW, LeGallo RD, Linden J, Kern JA, Tribble CG, Kron IL. Adenosine A2A receptor activation reduces inflammation and preserves pulmonary function in an in vivo model of lung transplantation. *J. Thorac. Cardiovasc. Surg.,* 129:1137-1143, 2005.

[51] Emaminia A, Lapar DJ, Zhao Y, Steidle JF, Harris DA, Laubach VE, Linden J, Kron IL, Lau CL. Adenosine A(2)A agonist improves lung function during ex vivo lung perfusion. *Ann. Thorac. Surg.,* 92:1840-1846, 2011.

[52] Hasko G, Xu DZ, Lu Q, Nemeth ZH, Jabush J, Berezina TL, Zaets SB, Csoka B, Deitch EA. Adenosine A2A receptor activation reduces lung injury in trauma/hemorrhagic shock. *Crit. Care Med.,* 34:1119-1125, 2006.

[53] Chen CM, Penuelas O, Quinn K, Cheng KC, Li CF, Zhang H, Slutsky AS. Protective effects of adenosine A2A receptor agonist in ventilator-induced lung injury in rats. *Crit. Care Med.,* 37:2235-2241, 2009.

[54] Reutershan J, Cagnina RE, Chang D, Linden J, Ley K. Therapeutic anti-inflammatory effects of myeloid cell adenosine receptor A2a stimulation in lipopolysaccharide-induced lung injury. *J. Immunol.,* 179:1254-1263, 2007.

[55] Field JJ, Lin G, Okam MM, Majerus E, Keefer J, Onyekwere O, Ross A, Campigotto F, Neuberg D, Linden J, Nathan DG. Sickle cell vaso-occlusion causes activation of iNKT cells that is decreased by the adenosine A2A receptor agonist regadenoson. *Blood,* 121:3329-3334, 2013.

[56] Schepp CP, Reutershan J. Bench-to-bedside review: adenosine receptors--promising targets in acute lung injury? *Crit. Care,* 12:226, 2008.

[57] Thadhani R, Pascual M, Bonventre JV. Acute renal failure. *N. Engl. J. Med.,* 334:1448-1460, 1996.

[58] Schroppel B, Legendre C. Delayed kidney graft function: from mechanism to translation. *Kidney Int.*, 86:251-258, 2014.

[59] Bellomo R, Kellum JA, Ronco C. Acute kidney injury. *Lancet,* 380:756-766, 2012.

[60] Liangos O, Wald R, O'Bell JW, Price L, Pereira BJ, Jaber BL. Epidemiology and outcomes of acute renal failure in hospitalized patients: a national survey. *Clin. J. Am. Soc. Nephrol.*, 1:43-51, 2006.

[61] Ishani A, Xue JL, Himmelfarb J, Eggers PW, Kimmel PL, Molitoris BA, Collins AJ. Acute kidney injury increases risk of ESRD among elderly. *J. Am. Soc. Nephrol.,* 20:223-228, 2009.

[62] Chawla LS, Kimmel PL. Acute kidney injury and chronic kidney disease: an integrated clinical syndrome. *Kidney Int.,* 82:516-524, 2012.

[63] Kinsey GR, Okusa MD. Pathogenesis of acute kidney injury: foundation for clinical practice. *Am. J. Kidney Dis.,* 58:291-301, 2011.

[64] Bauerle JD, Grenz A, Kim JH, Lee HT, Eltzschig HK. Adenosine generation and signaling during acute kidney injury. *J. Am. Soc. Nephrol.,* 22:14-20, 2011.

[65] Grenz A, Osswald H, Eckle T, Yang D, Zhang H, Tran ZV, Klingel K, Ravid K, Eltzschig HK. The reno-vascular A2B adenosine receptor protects the kidney from ischemia. *PLoS Med.,* 5:e137, 2008.

[66] Joo JD, Kim M, Horst P, Kim J, D'Agati VD, Emala CW, Sr., Lee HT. Acute and delayed renal protection against renal ischemia and reperfusion injury with A1 adenosine receptors. *Am. J. Physiol. Renal. Physiol.,* 293:F1847-1857, 2007.

[67] Okusa MD, Linden J, Macdonald T, Huang L. Selective A2A adenosine receptor activation reduces ischemia-reperfusion injury in rat kidney. *Am. J. Physiol.,* 277:F404-412, 1999.

[68] Lee HT, Emala CW. Protective effects of renal ischemic preconditioning and adenosine pretreatment: role of A(1) and A(3) receptors. *Am. J. Physiol. Renal. Physiol.,* 278:F380-387, 2000.

[69] Spielman WS, Arend LJ. Adenosine receptors and signaling in the kidney. *Hypertension,* 17:117-130, 1991.

[70] Levens N, Beil M, Jarvis M. Renal actions of a new adenosine agonist, CGS 21680A selective for the A2 receptor. *J. Pharmacol. Exp. Ther.,* 257:1005-1012, 1991.

[71] Silldorff EP, Kreisberg MS, Pallone TL. Adenosine modulates vasomotor tone in outer medullary descending vasa recta of the rat. *J. Clin. Invest.,* 98:18-23, 1996.

[72] Okusa MD, Linden J, Huang L, Rieger JM, Macdonald TL, Huynh LP. A(2A) adenosine receptor-mediated inhibition of renal injury and neutrophil adhesion. *Am. J. Physiol. Renal. Physiol.,* 279:F809-818, 2000.

[73] Day YJ, Huang L, Ye H, Li L, Linden J, Okusa MD. Renal ischemia-reperfusion injury and adenosine 2A receptor-mediated tissue protection: the role of CD4+ T cells and IFN-gamma. *J. Immunol.,* 176:3108-3114, 2006.

[74] Kinsey GR, Sharma R, Huang L, Li L, Vergis AL, Ye H, Ju ST, Okusa MD. Regulatory T cells suppress innate immunity in kidney ischemia-reperfusion injury. *J. Am. Soc. Nephrol.,* 20:1744-1753, 2009.

[75] Fontenot JD, Gavin MA, Rudensky AY. Foxp3 programs the development and function of CD4+CD25+ regulatory T cells. *Nat. Immunol.,* 4:330-336, 2003.

[76] Hori S, Nomura T, Sakaguchi S. Control of regulatory T cell development by the transcription factor Foxp3. *Science,* 299:1057-1061, 2003.

[77] Khattri R, Cox T, Yasayko SA, Ramsdell F. An essential role for Scurfin in CD4+CD25+ T regulatory cells. *Nat. Immunol.,* 4:337-342, 2003.

[78] Deaglio S, Dwyer KM, Gao W, Friedman D, Usheva A, Erat A, Chen JF, Enjyoji K, Linden J, Oukka M, Kuchroo VK, Strom TB, Robson SC. Adenosine generation catalyzed by CD39 and CD73 expressed on regulatory T cells mediates immune suppression. *J. Exp. Med.,* 204:1257-1265, 2007.

[79] Kinsey GR, Huang L, Vergis AL, Li L, Okusa MD. Regulatory T cells contribute to the protective effect of ischemic preconditioning in the kidney. *Kidney Int.,* 77:771-780, 2010.

[80] Kinsey GR, Huang L, Jaworska K, Khutsishvili K, Becker DA, Ye H, Lobo PI, Okusa MD. Autocrine adenosine signaling promotes regulatory T cell-mediated renal protection. *J. Am. Soc. Nephrol.,* 23:1528-1537, 2012.

[81] Li L, Huang L, Sung SS, Lobo PI, Brown MG, Gregg RK, Engelhard VH, Okusa MD. NKT cell activation mediates neutrophil IFN-gamma production and renal ischemia-reperfusion injury. *J. Immunol.*, 178: 5899- 5911, 2007.

[82] Yang SH, Lee JP, Jang HR, Cha RH, Han SS, Jeon US, Kim DK, Song J, Lee DS, Kim YS. Sulfatide-reactive natural killer T cells abrogate ischemia-reperfusion injury. *J. Am. Soc. Nephrol.,* 22:1305-1314, 2011.

[83] Li L, Okusa MD. Macrophages, dendritic cells, and kidney ischemia-reperfusion injury. *Semin. Nephrol.*, 30:268-277, 2010.

[84] Li L, Huang L, Sung SS, Vergis AL, Rosin DL, Rose CE, Jr., Lobo PI, Okusa MD. The chemokine receptors CCR2 and CX3CR1 mediate monocyte/macrophage trafficking in kidney ischemia-reperfusion injury. *Kidney Int.,* 74:1526-1537, 2008.

[85] Soos TJ, Sims TN, Barisoni L, Lin K, Littman DR, Dustin ML, Nelson PJ. CX3CR1+ interstitial dendritic cells form a contiguous network throughout the entire kidney. *Kidney Int.,* 70:591-596, 2006.

[86] Dong X, Swaminathan S, Bachman LA, Croatt AJ, Nath KA, Griffin MD. Resident dendritic cells are the predominant TNF-secreting cell in early renal ischemia-reperfusion injury. *Kidney Int.,* 71:619-628, 2007.

[87] Bullough DA, Magill MJ, Firestein GS, Mullane KM. Adenosine activates A2 receptors to inhibit neutrophil adhesion and injury to isolated cardiac myocytes. *J. Immunol.,* 155:2579-2586, 1995.

[88] Cronstein BN, Levin RI, Philips M, Hirschhorn R, Abramson SB, Weissmann G. Neutrophil adherence to endothelium is enhanced via adenosine A1 receptors and inhibited via adenosine A2 receptors. *J. Immunol.,* 148:2201-2206, 1992.

[89] Kelly KJ, Williams WW, Jr., Colvin RB, Meehan SM, Springer TA, Gutierrez-Ramos JC, Bonventre JV. Intercellular adhesion molecule-1-deficient mice are protected against ischemic renal injury. *J. Clin. Invest.,* 97:1056-1063, 1996.

[90] Awad AS, Rouse M, Huang L, Vergis AL, Reutershan J, Cathro HP, Linden J, Okusa MD. Compartmentalization of neutrophils in the kidney and lung following acute ischemic kidney injury. *Kidney Int.,* 75:689-698, 2009.

[91] Day YJ, Marshall MA, Huang L, McDuffie MJ, Okusa MD, Linden J. Protection from ischemic liver injury by activation of A2A adenosine receptors during reperfusion: inhibition of chemokine induction. *Am. J. Physiol. Gastrointest Liver Physiol.,* 286:G285-293, 2004.

[92] Frangogiannis NG. The inflammatory response in myocardial injury, repair, and remodelling. *Nat. Rev. Cardiol.,* 11:255-265, 2014.

[93] Olafsson B, Forman MB, Puett DW, Pou A, Cates CU, Friesinger GC, Virmani R. Reduction of reperfusion injury in the canine preparation by intracoronary adenosine: importance of the endothelium and the no-reflow phenomenon. *Circulation,* 76:1135-1145, 1987.

[94] Linden J. Molecular approach to adenosine receptors: receptor-mediated mechanisms of tissue protection. *Annu. Rev. Pharmacol. Toxicol.,* 41:775-787, 2001.

[95] Thornton JD, Liu GS, Olsson RA, Downey JM. Intravenous pretreatment with A1-selective adenosine analogues protects the heart against infarction. *Circulation,* 85:659-665, 1992.

[96] Zhao ZQ, Sato H, Williams MW, Fernandez AZ, Vinten-Johansen J. Adenosine A2-receptor activation inhibits neutrophil-mediated injury to coronary endothelium. *Am. J. Physiol. Heart Circ. Physiol.,* 271:H1456-1464, 1996.

[97] Jordan JE, Thourani VH, Auchampach JA, Robinson JA, Wang N-P, Vinten-Johansen J. A3 adenosine receptor activation attenuates neutrophil function and neutrophil-mediated reperfusion injury. *Am. J. Physiol. Heart Circ. Physiol.,* 277:H1895-1905, 1999.

[98] Philipp S, Yang X-M, Cui L, Davis AM, Downey JM, Cohen MV. Postconditioning protects rabbit hearts through a protein kinase C-adenosine A2b receptor cascade. *Cardiovascular Research,* 70:308-314, 2006.

[99] Jordan JE, Zhao ZQ, Sato H, Taft S, Vinten-Johansen J. Adenosine A2 receptor activation attenuates reperfusion injury by inhibiting neutrophil accumulation, superoxide generation and coronary endothelial adherence. *J. Pharmacol. Exp. Ther.,* 280:301-309, 1997.

[100] Lasley RD, Jahania MS, Mentzer RM, Jr. Beneficial effects of adenosine A2a agonist CGS-21680 in infarcted and stunned porcine myocardium. *Am. J. Physiol. Heart Circ. Physiol.,* 280:H1660-1666, 2001.

[101] Glover DK, Riou LM, Ruiz M, Sullivan GW, Linden J, Rieger JM, Macdonald TL, Watson DD, Beller GA. Reduction of infarct size and postischemic inflammation from ATL-146e, a highly selective adenosine A2A receptor agonist, in reperfused canine myocardium. *Am. J. Physiol. Heart Circ. Physiol.,* 288:H1851-1858, 2005.

[102] Schlack W, Schafer M, Uebing A, Schafer S, Borchard U, Thamer V. Adenosine-A2-receptor activation at reperfusion reduces infarct size and improves myocardial wall function in dog heart. *J. Cardiovasc. Pharmacol.,* 22:89-96, 1993.

[103] Norton ED, Jackson EK, Turner MB, Virmani R, Forman MB. The effects of intravenous infusions of selective adenosine-A1-receptor and A2-receptor agonists on myocardial reperfusion injury. *Am. Heart J.,* 123:332-338, 1992.

[104] Dai W, Hale SL, Nayak R, Kloner RA. ATL 313, a selective A(2A) adenosine receptor agonist, reduces myocardial infarct size in a rat ischemia/reperfusion model. *Open. Cardiovasc. Med. J.,* 3:166-172, 2009.

[105] Yang Z, Day Y-J, Toufektsian M-C, Ramos SI, Marshall M, Wang X-Q, French BA, Linden J. Infarct-sparing effect of A2A-adenosine receptor activation is due primarily to its action on lymphocytes. *Circulation,* 111:2190-2197, 2005.

[106] Zhao ZQ, Corvera JS, Halkos ME, Kerendi F, Wang NP, Guyton RA, Vinter-Johansen J. Inhibition of myocardial injury by ischemic postconditioning during reperfusion: comparison with ischemic preconditioning. *Am. J. Physiol. Heart Circ. Physiol.,* 285:H579-H588, 2003.

[107] Granfeldt A, Lefer DJ, Vinten-Johansen J. Protective ischaemia in patients: preconditioning and postconditioning. *Cardiovasc. Res.,* 83:234-246, 2009.

[108] Morrison RR, Tan XL, Ledent C, Mustafa SJ, Hofmann PA. Targeted deletion of A2A adenosine receptors attenuates the protective effects of myocardial postconditioning. *Am. J. Physiol Heart Circ. Physiol.,* 293:H2523-2529, 2007.

[109] Massip-Salcedo M, Rosello-Catafau J, Prieto J, Avila MA, Peralta C. The response of the hepatocyte to ischemia. *Liver Int.,* 27:6-16, 2007.

[110] Ben-Ari Z, Pappo O, Sulkes J, Cheporko Y, Vidne BA, Hochhauser E. Effect of adenosine A2A receptor agonist (CGS) on ischemia/reperfusion injury in isolated rat liver. *Apoptosis,* 10:955-962, 2005.

[111] Harada N, Okajima K, Murakami K, Usune S, Sato C, Ohshima K, Katsuragi T. Adenosine and selective A(2A) receptor agonists reduce ischemia/reperfusion injury of rat liver mainly by inhibiting leukocyte activation. *J Pharmacol. Exp. Ther.,* 294:1034-1042, 2000.

[112] Gantner F, Leist M, Lohse AW, Germann PG, Tiegs G. Concanavalin A-induced T-cell-mediated hepatic injury in mice: the role of tumor necrosis factor. *Hepatology,* 21:190-198, 1995.

[113] Odashima M, Otaka M, Jin M, Horikawa Y, Matsuhashi T, Ohba R, Linden J, Watanabe S. A selective adenosine A2A receptor agonist, ATL-146e, prevents concanavalin A-induced acute liver injury in mice. *Biochem. Biophys. Res. Commun.,* 347:949-954, 2006.

[114] Ohta A, Sitkovsky M. Role of G-protein-coupled adenosine receptors in downregulation of inflammation and protection from tissue damage. *Nature,* 414:916-920, 2001.

[115] Chouker A, Thiel M, Lukashev D, Ward JM, Kaufmann I, Apasov S, Sitkovsky MV, Ohta A. Critical role of hypoxia and A2A adenosine receptors in liver tissue-protecting physiological anti-inflammatory pathway. *Mol. Med.,* 14:116-123, 2008.

[116] Cao Z, Yuan Y, Jeyabalan G, Du Q, Tsung A, Geller DA, Billiar TR. Preactivation of NKT cells with alpha-GalCer protects against hepatic ischemia-reperfusion injury in mouse by a mechanism involving IL-13 and adenosine A2A receptor. *Am. J. Physiol. Gastrointest Liver Physiol.,* 297:G249-258, 2009.

[117] Day YJ, Li Y, Rieger JM, Ramos SI, Okusa MD, Linden J. A2A adenosine receptors on bone marrow-derived cells protect liver from ischemia-reperfusion injury. *J. Immunol.,* 174:5040-5046, 2005.

[118] Hasko G, Kuhel DG, Chen JF, Schwarzschild MA, Deitch EA, Mabley JG, Marton A, Szabo C. Adenosine inhibits IL-12 and TNF-[alpha] production via adenosine A2a receptor-dependent and independent mechanisms. *FASEB J.,* 14:2065-2074, 2000.

[119] Tang LM, Zhu JF, Wang F, Qian J, Zhu J, Mo Q, Lu HH, Li GQ, Wang XH. Activation of adenosine A2A receptor attenuates inflammatory response in a rat model of small-for-size liver transplantation. *Transplant. Proc.,* 42:1915-1920, 2010.

[120] Nowak M, Lynch L, Yue S, Ohta A, Sitkovsky M, Balk SP, Exley MA. The A2aR adenosine receptor controls cytokine production in iNKT cells. *Eur. J. Immunol.,* 40:682-687, 2010.

[121] Alchera E, Tacchini L, Imarisio C, Dal Ponte C, De Ponti C, Gammella E, Cairo G, Albano E, Carini R. Adenosine-dependent activation of hypoxia-inducible factor-1 induces late preconditioning in liver cells. *Hepatology,* 48:230-239, 2008.

[122] Chouker A, Ohta A, Martignoni A, Lukashev D, Zacharia LC, Jackson EK, Schnermann J, Ward JM, Kaufmann I, Klaunberg B, Sitkovsky MV, Thiel M. In vivo hypoxic preconditioning protects from warm liver ischemia-reperfusion injury through the adenosine A2B receptor. *Transplantation,* 94:894-902, 2012.

[123] Sakowicz-Burkiewicz M, Kocbuch K, Grden M, Szutowicz A, Pawelczyk T. Diabetes-induced decrease of adenosine kinase expression impairs the proliferation potential of diabetic rat T lymphocytes. *Immunology,* 118:402-412, 2006.

[124] Celinski K, Prozorow-Krol B, Korolczuk A, Slomka M, Korobowicz E, Biskup W, Madro A, Cichoz-Lach H, Czechowska G. The role of adenosine A2a receptors in experimental acute pancreatitis. *Adv. Med. Sci.,* 51:105-110, 2006.

[125] Awad AS, Huang L, Ye H, Duong ET, Bolton WK, Linden J, Okusa MD. Adenosine A2A receptor activation attenuates inflammation and injury in diabetic nephropathy. *Am. J. Physiol. Renal. Physiol.,* 290:F828-837, 2006.

[126] Nemeth ZH, Bleich D, Csoka B, Pacher P, Mabley JG, Himer L, Vizi ES, Deitch EA, Szabo C, Cronstein BN, Hasko G. Adenosine receptor activation ameliorates type 1 diabetes. *FASEB J.,* 21:2379-2388, 2007.

[127] Zarek PE, Huang CT, Lutz ER, Kowalski J, Horton MR, Linden J, Drake CG, Powell JD. A2A receptor signaling promotes peripheral tolerance by inducing T-cell anergy and the generation of adaptive regulatory T cells. *Blood,* 111:251-259, 2008.

[128] Nitta T, Itoh T, Matsuoka N, Mera T, Kojima D, Nakano M, Yamashita Y, Yasunami Y. Prevention of early loss of transplanted islets in the liver of mice by adenosine. *Transplantation,* 88:49-56, 2009.

[129] Toyofuku A, Yasunami Y, Nabeyama K, Nakano M, Satoh M, Matsuoka N, Ono J, Nakayama T, Taniguchi M, Tanaka M, Ikeda S. Natural killer T-cells participate in rejection of islet allografts in the liver of mice. *Diabetes,* 55:34-39, 2006.

[130] Yasunami Y, Kojo S, Kitamura H, Toyofuku A, Satoh M, Nakano M, Nabeyama K, Nakamura Y, Matsuoka N, Ikeda S, Tanaka M, Ono J, Nagata N, Ohara O, Taniguchi M. Valpha14 NK T cell-triggered IFN-gamma production by Gr-1+CD11b+ cells mediates early graft loss of syngeneic transplanted islets. *J. Exp. Med.,* 202:913-918, 2005.

[131] Ghaemi Oskouie F, Shameli A, Yang A, Desrosiers MD, Mucsi AD, Blackburn MR, Yang Y, Santamaria P, Shi Y. High levels of adenosine deaminase on dendritic cells promote autoreactive T cell activation and diabetes in nonobese diabetic mice. *J. Immunol.,* 186:6798-6806, 2011.

[132] Chia J, Robson S, Mysore T, d'Apice A, Cowan P, Dwyer K. Expression of human CD39 on mouse islets protects against chemically induced diabetes (abstract). *Xenotransplantation,* 16:328, 2009.

[133] Chhabra P, Linden J, Lobo P, Douglas Okusa M, Lewis Brayman K. The immunosuppressive role of adenosine A2A receptors in ischemia reperfusion injury and islet transplantation. *Curr. Diabetes. Rev.,* 8:419-433, 2012.

[134] Chhabra P, Wang K, Zeng Q, Jecmenica M, Langman L, Linden J, Ketchum RJ, Brayman KL. Adenosine A(2A) agonist administration improves islet transplant outcome: Evidence for the role of innate immunity in islet graft rejection. *Cell Transplant,* 19:597-612, 2010.

[135] Luijk B, van den Berge M, Kerstjens HA, Postma DS, Cass L, Sabin A, Lammers JW. Effect of an inhaled adenosine A2A agonist on the allergen-induced late asthmatic response. *Allergy,* 63:75-80, 2008.

[136] Mantell S, Jones R, Trevethick M. Design and application of locally delivered agonists of the adenosine A(2A) receptor. *Expert. Rev. Clin. Pharmacol.,* 3:55-72, 2010.

[137] Murphree LJ, Marshall MA, Rieger JM, MacDonald TL, Linden J. Human A(2A) adenosine receptors: high-affinity agonist binding to receptor-G protein complexes containing Gbeta(4). *Mol. Pharmacol.,* 61:455-462, 2002.

[138] Bonneau O, Wyss D, Ferretti S, Blaydon C, Stevenson CS, Trifilieff A. Effect of adenosine A2A receptor activation in murine models of respiratory disorders. *Am. J. Physiol. Lung Cell Mol. Physiol.,* 290: L1036- 1043, 2006.

[139] Fozard JR, Ellis KM, Villela Dantas MF, Tigani B, Mazzoni L. Effects of CGS 21680, a selective adenosine A2A receptor agonist, on allergic airways inflammation in the rat. *Eur. J. Pharmacol.,* 438:183-188, 2002.

[140] Moore CC, Martin EN, Lee GH, Obrig T, Linden J, Scheld WM. An A2A adenosine receptor agonist, ATL313, reduces inflammation and improves survival in murine sepsis models. *BMC Infect. Dis.,* 8:141, 2008.

[141] Warren CA, Calabrese GM, Li Y, Pawlowski SW, Figler RA, Rieger J, Ernst PB, Linden J, Guerrant RL. Effects of adenosine A(2)A receptor activation and alanyl-glutamine in Clostridium difficile toxin-induced ileitis in rabbits and cecitis in mice. *BMC Infect. Dis.,* 12:13, 2012.

In: Adenosine Receptors
Editor: Kasandra Warrick
ISBN: 978-1-63463-454-0

Chapter II

Adenosine Receptors in the Vasculature: Physiology, Pharmacology and Therapeutic Perspectives

***J. B. Sousa*[1,2], *A. S. Soares*[1,2], *P. Fresco*[1,2] and *C. Diniz*[1,2]**
[1]REQUIMTE/FARMA, Department of Drug Sciences, Laboratory of Pharmacology, University of Porto, Porto, Portugal
[2]MedinUP- Center for Drug Discovery and Innovative Medicines, University of Porto, Porto, Portugal

Abstract

Adenosine is a ubiquitous adenine base purine that mediates diverse biological and physiological processes in the cardiovascular system, both in cardiac and vascular tissues. Adenosine was reported to be involved in smooth muscle contraction, neurotransmission/neuromodulation, immune response and inflammation, platelet aggregation and modulation of cardiac function.

Adenosine availability, in the extracellular space, depends on its release from a variety of cell types, as a result of adenosine triphosphate (ATP) enzymatic breakdown of both intra- and extracellular adenine nucleotides and intracellular S-adenosylhomocysteine. Other important source of extracellular adenosine lies on the activity of nucleoside transporters, both equilibrative (ENT) and concentrative (CNT), particularly ENT1, ENT4 and CNT2. As such adenosine can reach concentrations able to activate adenosine receptors. Four adenosine receptor subtypes have been identified, A_1, A_{2A}, A_{2B} and A_3 receptors, and were found to trigger different signalling pathways and to present differential affinity requirements.

Correlation between the ENT activity and adenosine receptor activation is also discussed focusing on the impact that changes or disruption of the interplay between these effectors may cause in the vascular physiology, namely by contributing to disease.

All adenosine receptor subtypes have been identified in the vascular wall of many different arteries and veins. A brief overview of the current knowledge and of our research work will be integrated considering important aspects of the adenosine-vascular interaction with particular emphasis on adenosine receptor subtypes pharmacology in the three vascular layers, intima, media and adventitia, namely by explaining their individual contribution for vasoconstriction/vasodilation. Integration of these knowledge with the morphological features of arteries/veins, namely by evidencing how innervation may vary from dense to sparse (depending on vessel localization) or how smooth muscle cell layers of the media varies (depending on vessel type and/or localization) will also be discussed, clarifying the contribution of adenosine receptors to peripheral vascular resistance.

In the last decade several new adenosine receptor ligands have been identified as having promising therapeutic activities but during clinical trials some of them have been discontinued due to side effects. Despite of this fact, a renewed interest in this research area is emerging and several new adenosine receptor ligands have been developed and are undergoing clinical trials for a broad range of indications such as neurodegenerative disorders, inflammation, wound healing, cardiac imaging, etc. In summary, in this chapter an overview of adenosine/adenosine receptors physiology and pharmacology in the vasculature is presented guiding the reader to new and putative therapeutic applications.

1. Introduction

Adenosine mediates diverse biological and physiological processes in the cardiovascular system, both in cardiac and vascular tissues [1]. Adenosine was reported to be involved in smooth muscle contraction, neurotransmission/ neuromodulation, immune response and inflammation, platelet aggregation and modulation of cardiac function. Indeed, adenosine appears to be the direct modulator of both cardiac and vascular functions, but may also have indirect effects by virtue of its influence on the autonomic nervous system, mast cells and platelet aggregation. It is also recognized that adenosine actions in vessels are more prominent in hypoxic conditions where high levels of adenosine are present. Adenosine levels seem to be also higher in several pathological conditions revealing the importance of adenosine and adenosine receptors as therapeutical targets.

2. Adenosine Bioavailability

Adenosine is an adenine base presenting two carbon-nitrogen rings, bound to a ribose sugar group *via* a beta glycosidic link. Adenosine presents a short half-life due to its rapid conversion into inosine by adenosine deaminase, phoshorylation by adenosine kinase and uptake by adenosine transporters.

2.1. Adenosine Formation and Catabolism

There are several pools of adenosine which arise from different sources. Adenosine availability, in the extracellular space, depends on its release from a variety of cell types, as a result of adenosine triphosphate (ATP) enzymatic breakdown of both intra- and extracellular adenine nucleotides and intracellular S-adenosylhomocysteine. ATP present in the cytosol can be dephosphorylated to adenosine monophosphate (AMP) and, then, further dephosphorylated by the action of adenosine kinase producing adenosine. Alternatively, ATP can be released from the cell by exocytosis, which can then be metabolized by nucleotidases to form adenosine diphosphate (ADP), then AMP and, finally, adenosine.

Under physiological conditions, adenosine is mainly produced intracellularly by AMP dephosphorylation, and extracellularly by dephosphorylation of released adenine nucleotides, mainly ATP [2, 3]. In the extracellular environment, adenosine concentrations can be regulated by different enzyme pathways: cyclic adenosine monophosphate (cAMP), produced intracellular upon adenylyl cyclase activation, can be transported out of cells where upon ecto-phosphodiesterase activity AMP is formed or, in some cells types, AMP can be directly released out of the cell; AMP is, then, dephosphorylated by an ecto-5'-nucleotidase to form adenosine. In the intracellular environment, adenosine concentrations can be regulated by other enzyme pathways: AMP can be directly deaminated to inosine monophosphate (IMP) by AMP deaminase, or acted upon by an endo-5'-nucleotidase to form adenosine; it can also be rephosphorylated to AMP by adenosine kinase, or deaminated to inosine by adenosine deaminase. IMP can also be a source of inosine by the same endo-5'-nucleotidase. Hypoxanthine is formed after removal of ribose from inosine by the actions of purine nucleoside phosphorylase. Hypoxanthine can be salvaged to IMP by hypoxanthine-phosphoribosyltransferase, or by entering the xanthine oxidase pathway to sequentially form xanthine and uric acid, generating oxyradicals.

Intracellular adenosine can be transported into and out of the cell by membrane-associated transporter proteins.

2.2. Adenosine Transport

Once it is formed adenosine can then be transported between the inside of the cell and the interstitial fluid or *vice-versa, via* transporters. Indeed, adenosine transport proteins can move adenosine across cellular membranes, influencing extracellular adenosine concentrations [2]. Transport of adenosine occurs through a family of membrane proteins that present different pharmacological and kinetic properties [4], the nucleoside transporters. These membrane proteins have been recently identified and characterized in humans and were purified from human blood red cells for more than two decades ago. Indeed, the lack of abundance of nucleoside transporter proteins in mammals has hampered the study of its structure and function [5-7].

It is well accepted that nucleoside transporters facilitate the movement of nucleosides and nucleobases across cell membranes. However, nucleoside transporters are not homogeneously distributed among tissues and their expression seems to be regulated by several physiological and pathophysiological conditions [7-9]. In the last twenty years important findings have helped the understanding of nucleoside transporters function. Nucleoside transporters can i) salvage extracellular nucleosides for intracellular synthesis of nucleotides, ii) control the extracellular concentration of adenosine in the vicinity of its cell surface receptors, iii) help to regulate processes such as neurotransmission and cardiovascular activity [10, 11] and iv) modulate cell function since, they supply vital nucleosides to the synthesis of nucleic acids in cells that lack *de novo* purine synthesis. Therefore, the efficiency of the transport processes determines adenosine availability to its receptors or to metabolizing enzymes. Nucleoside transporters can be considered promising therapeutic targets in the cardiovascular disorders as they have, for example, been used in anticancer and antiviral therapy, where nucleoside analogues are, currently, used successfully [9, 12-14].

Currently, two types of nucleoside transporters have been identified [8, 15]: Equilibrative Nucleoside Transporters (ENT) and Concentrative Transporters (CNT) [16].

2.2.1. Equilibrative Nucleoside Transporters

Equilibrative Nucleoside Transporters mediate equilibrative bidirectional transport processes driven by chemical gradients by facilitated diffusion.

A decade ago, the first example of the ENT family, in human tissues, was characterized at the molecular level and, since then, identification of homologous proteins by functional cloning and genome analysis has revealed that this family is widely distributed in eukaryotes, the SLC29 family. Studies have revealed that most mammalian cells exhibit low-affinity, ENT processes, now known to be mediated by members of the SLC29 family. ENT has no apparent sequence homology to other types of transporters, which enable facilitated diffusion of nucleosides, such as adenosine, and nucleoside analogues across cell membranes [17].

ENTs are present in most, possibly all, cell types [18]. ENT family members are believed to possess eleven transmembrane domains, with a cytoplasmic N-terminus and an extracellular C-terminus (confirmed for ENT1: [8]). The number of molecules present of each ENT subtype depends on both the cell and the tissue type.

ENT might mediate adenosine transport in both directions, depending on the concentration gradient of adenosine across the plasma membrane. Until the present day, there are four subtypes described: ENT1, ENT2, ENT3 and ENT4 [6, 8, 15]. The best characterized ENT, ENT1 and ENT2, are cell surface proteins possessing similar broad substrate specificities for both purine and pyrimidine nucleosides and can regulate the access of adenosine to its receptors. Indeed, ENT1 seem to play a primary role mediating adenosine transport whereas ENT2, additionally, transport nucleobases. ENT3 and ENT4 isoforms have been shown to present ability as genuine nucleoside transporters, although their optimally active conditions differ from ENT1 and ENT2: they are pH sensitive being active under acidic conditions. ENT3 has a similar broad selectivity for nucleosides and nucleobases and it was identified in intracellular membranes, including lysosomes (with preserved transport ability). ENT4 present low affinity for adenosine yet is selective for adenosine and for several organic cations [8, 19].

All ENT members share the ability to transport adenosine but their abilities to transport other nucleosides and nucleobases differ. The four isoforms are widely distributed in mammalian tissues. Nevertheless, ENT1, ENT2, ENT3 and ENT4 relative abundance varies. In polarized cells, ENT, particularly ENT1 and ENT2, are found in the basolateral membrane and in tandem with other members of SLC28 family (such as CNT), may be involved in transepithelial nucleoside transport. ENT2 is known to be particularly abundant in skeletal muscle. ENT3 and ENT4 isoforms seem to be widely distributed with ENT3 being the most abundant ENT in the heart and with ENT4 being present in vascular endothelial cells. Nevertheless, since ENT3 is a lysosomal nucleoside transporter being active in intracellular membranes, most probably is devoid of effect in a direct regulation of tissue interstitial adenosine concentrations. ENT4 presents low sequence identity to other SLC29 members of the family (mainly due to structural differences), is, however, highly selective for adenosine [8, 20].

2.2.2. Concentrative Transporters

Concentrative Transporters mediate active inwardly directed concentrative processes, driven by the Na^+ electrochemical gradient, Na^+-dependent.

Identification and molecular cloning of CNT families, from mammals and protozoan parasites, have provided detailed information about tissue and cellular localization, the structure, function and regulation [7, 8]. CNT belong to the SLC28 family and are expressed in a tissue-specific fashion with CNT transport processes being present primarily in specialized epithelia [18]. CNT are tightly regulated, both by endocrine and growth factors and by substrate availability. They transport endogenous substrates such as adenosine, thymidine, guanosine, uridine, inosine as well as hypoxanthine [9]. They are involved in the transport of adenosine, although ENT has higher affinity for adenosine than CNT. Three subtypes were described: CNT1, CNT2 and CNT3 [6, 17, 19]. All CNTs serve to transport uridine but present differential affinities for other permeants: CNT1 is selective for pyrimidine nucleosides transport (including adenosine at lower fluxes); CNT2 present selectivity for purine nucleosides; CNT3 serve to transport both purine and pyrimidine nucleosides.

CNT3 is widely tissue distributed than the other CNTs [21] and may have diverse roles in intracellular compartments [22]. Considerable body of evidence as linked the reduction of expression of CNTs to resistance to therapeutic regimens, namely with anti-cancer drugs such as 5-fluorouracil, 6-mercaptopurine and 6-thioguanine [23, 24] which can be explained by the fact that CNTs serve as nucleoside and nucleoside drug import pumps [21, 25, 26]. Among the nucleosides transported by CNTs are anticancer and antiviral nucleoside drugs.

2.2.3. Adenosine Transport Inhibition

ENTs are sensitive to nitrobenzylthioinosine (NBMPR), and are also inhibited by coronary vasodilators such as dipyridamole, dilazep, and draflazine: ENT1> ENT2>ENT3>>ENT4 [8, 15, 27, 28]. Less potent inhibitors of ENT1 and ENT2 include tyrosine- and serine/threonine-kinase inhibitors and benzodiazepines [29]. Dipyridamole, dilazep, the organic cation transporter inhibitor decynium-22, and the dopamine transporter inhibitor GBR125935 are low-affinity inhibitors of ENT4 [20].

ENTs inhibitors have been extensively used in pharmacological research and, at present, are under clinical investigation to increase the therapeutic indices of both nucleoside and non-nucleoside antimetabolite drugs [30, 31]. Concerning CNTs inhibition, despite efforts in the past years, only recently some molecules have been identified as having the ability to function as tight-binding, selective and non-transportable inhibitors of CNT1. These molecules are fused pyrimidine nucleosides, thienopyrimidine 2′-deoxynucleoside (dMeThPmR) and ribonucleoside (MeThPmR) [32].

3. Adenosine Receptors

Reports of adenosine and AMP effects on heart and blood vessels were the first in a major line of research concerning the physiological actions of purines. [33] Since then, the list of biological processes in which extracellular purines participate has increased. Insights into the physiological roles of purines come from studies of their biological sources and the stimuli for their release. Adenosine involvement was reported in smooth muscle contraction, neurotransmission and neuromodulation, exocrine and endocrine secretion, immune response and inflammation, platelet aggregation, pain and modulation of cardiac function [34, 35].

Currently it is accepted that adenosine mediated actions occurs by interacting with integral membrane proteins, known as A_1, A_{2A}, A_{2B} and A_3. [36] The classification of adenosine receptors emerged after an intensive research of adenosine effects on adenylyl cyclase activity: inhibitory or stimulatory effects [37]. Inhibition of adenylyl cyclase by adenosine ligands requires an unmodified purine ring whereas adenylyl cyclase stimulation requires ligands with an intact ribose ring. These observations led to the proposal of two distinct adenosine "reactive sites", called P and R sites, respectively: the R site was assumed to be located at the membrane, a "G protein related class" whereas the P site was hypothesized to be intracellularly located [37]. According to the effects upon adenylyl cyclase adenosine receptors were further divided into adenosine A_1 and A_2 receptors [38], to designate the adenylyl cyclase inhibitory and stimulatory adenosine receptors, respectively.

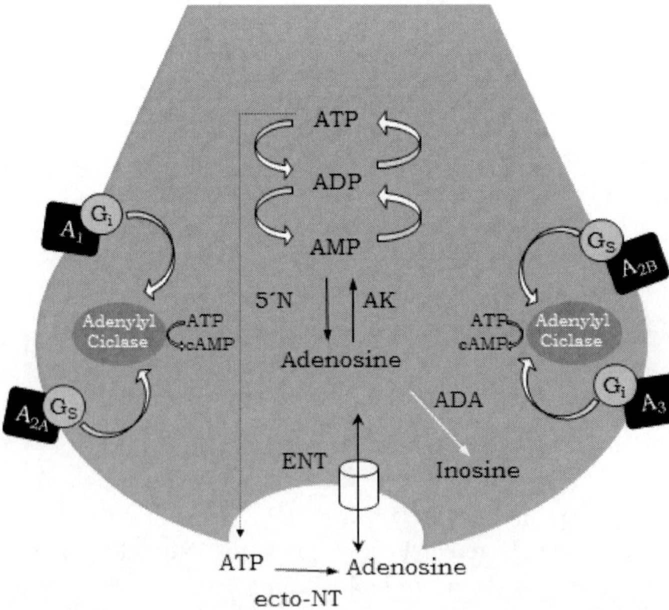

ATP – adenosine triphosphate; ADP – adenosine diphosphate; AMP – adenosine monophosphate;

5′-N – 5′-nucleotidase; AK – adenosine kinase; ADA – adenosine deaminase; ecto-NT – ecto-nucleotidase.

Figure 1. Schematic representation of adenosine formation, metabolism and transport, and G-protein coupled adenosine receptors (rectangles representing A_1, A_{2A}, A_{2B}, and A_3 receptors). Neurons can release adenosine and ATP. All cell types express adenosine receptors, adenosine transporters (cylinder), and ecto-nucleotidases that convert ATP into adenosine. Activation of adenosine receptors either inhibits ($A_{1/3}$ receptors) or stimulates ($A_{2A/2B}$ receptors) adenylyl cyclase and the cyclic AMP (cAMP) pathway.

An alternative nomenclature, Ri and Ra receptors, based in the same criteria, was proposed by Londos and coworkers (1980) [39]. The nomenclature adopted by the IUPHAR Nomenclature Committee (NC-IUPHAR) was the initially proposed designation of adenosine A_1 and A_2 receptors [40]. The subclassification of adenosine A_2 receptors was proposed at 1983 by Daly and co-workers [41] based on the ability of adenosine to stimulate adenylyl cyclase in brain slices: A_{2A} with high affinity for adenosine and A_{2B} with lower affinity.

The classification of adenosine receptor subtypes, largely based on the pharmacological profiles of A_1, A_{2A} and A_{2B} receptors, was confirmed after their cloning (A_1;[42] A_{2A};[43] A_{2B};[44]),[45]). A fourth subtype of adenosine receptor, A_3 remained pharmacologically undetected until its cloning [46, 47]. Adenosine A_3 receptors reveal differences relatively to the other adenosine receptor subtypes: sequence identity of the human subtype varies from 43% to 54%. Moreover, adenosine A_3 receptors present a remarkably low identity between species comparatively to those observed with the other adenosine receptor subtypes [48].

Adenosine A_1 and A_{2A} receptors are considered high affinity receptors for adenosine but present differences in their respective K_d with A_{2A} receptors requiring concentrations higher (1-20 nM) than those required to activate adenosine A_1 receptors (0.3-3 nM) [49]. Adenosine A_{2B} and A_3 receptors are considered low affinity receptors since they require high amounts of adenosine to be activated (>1 μM) [50]. Adenosine availability is, therefore, crucial to determine the subtype of adenosine receptors activated. Figure 1 represents the processes of adenosine synthesis, metabolism, transport and adenosine/G-protein mediated pathway.

3.1. Interplay between Nucleoside Transport and Adenosine Receptor Activity

It has been speculated that the increase in the activities of ENT1 and CNT2 may reduce the availability of adenosine to its receptors, conditioning its effects. For example, ENTs are likely to modulate adenosine-mediated effects in the sinoatrial node of the heart, as described by King and co-workers (2006). They found dipyridamole potentiates A_1 receptor-mediated chronotropic effects of adenosine *via* inhibition of uptake [30]. Another example was the demonstration, in human airway epithelial cells, where ENT1 and ENT2 were found to modulate adenosine-mediated effects of the cystic fibrosis transmembrane conductance regulator (CFTR) and K^+ channels. [8, 51] Other evidence come from *Slc29a1*-null mice studies that revealed an important role of ENT1 in i) ethanol preference and consumption [52-54] ii) anxiety-related behaviour [55, 56] and iii) cardioprotection during ischemia [57]. These changes can be explained by altered ENT1-mediated control of adenosine levels and adenosine receptor signaling and, consistent with this, *Slc29a1*-null mice have elevated plasma levels of adenosine [57].

In vascular tissues, recent studies concerning the role of endothelium in hypertension, suggest that the main mechanism regulating extracellular adenosine levels involves nucleoside uptake to endothelial cells with the subsequent impairment of adenosine A_1 receptor activation [58]. Thus, nucleoside transporters may play an integral part in adenosine functions by "fine-tuning" local levels of adenosine in the vicinity of adenosine receptors.

3.2. Molecular Structure and Ligand Binding of Adenosine Receptors

All adenosine receptors present seven transmembrane (TM I-VII) domains of hydrophobic amino acids, each believe to constitute a α-helice of approximately 21 to 28 amino acids. TM I-VII domains are connected by three extracellular and three intracellular hydrophilic and different size loops. The extracellular loop between TM IV and TM V as well as the cytoplasmatic loop between TM V and TM VI are extended [59]. TM I-VII regions are generally highly conserved and its residues are crucial pieces for ligand binding and specificity [60, 61]. The N-terminal of this integral protein lies on the extracellular side whereas the C-terminal lies on the cytoplasmatic side of the plasma membrane. (Figure 2 represents a schematic of the A_1 adenosine receptor).

The ligand binding site is a "pocket" formed 3D-arrangement of the transmembrane domains. The most essential interactions required for ligand recognition occur in TM III, V, VI, VII. Two histidine residues are conserved among adenosine A_1, A_{2A} and A_{2B} receptors at TM VI, position 52. These two histidine residues are crucial for ligand recognition [63-65] contributing to ligand specificity within the binding pocket. Adenosine A_3 receptors constitution lacks these two histidine residues (TM VI: His52). Other amino acids residues seem to have a role in ligand recognition, affinity and binding [64, 66-69]. As an example, the glutamate in TM I of the human adenosine

A_{2A} receptor seem to be critically involved in agonist, but not in antagonist, recognition [70].

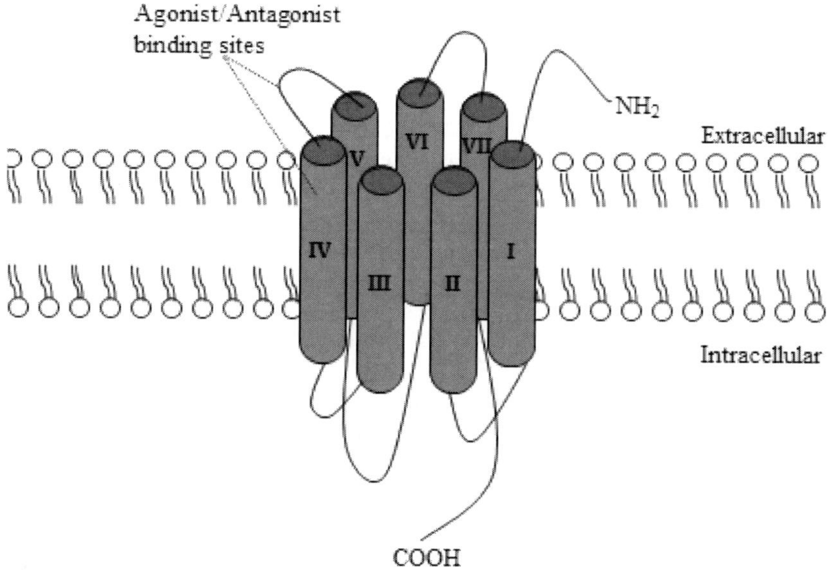

Figure 2. Schematic of the A_1 adenosine receptor. A_1 receptor has seven putative transmembrane domains (I-VII) of hydrophobic amino acids, each believed to constitute a α-helix, which are connected by three extracellular and three intracellular hydrophilic loops. Extracellular and transmembrane regions of the protein believed to be important in agonist and antagonist binding are indicated [60, 62].

The distal region of the second extracellular loop of adenosine receptors has also a role in the binding of both agonists and antagonists [61]: it requires residues negatively charged in adenosine A_{2A} receptor for the binding to both agonists and antagonists. Concerning adenosine A_{2A} receptors, site-directed mutagenesis, has identified several other residues in TM V, VI, VII involved in ligand binding [64]. For example, a dramatic loss of affinity for ligands was found when histidine residues at TM VI, position 52 and TM VII, position 43 were mutated at adenosine A_{2A} receptor; substitution of histidine in TM VII, position 43 for other residues also reduced agonist affinity [65].

Also important in ligand binding are the cysteine residues that form a disulfide bridge in the third extracellular loop. These residues are conserved among rhodopsin-like G-protein couple receptor family, being required for structural integrity of the receptor and also essential for ligand binding [71].

Finally, mutations in TM IV in adenosine receptors ruled out its influence in ligand binding [72]. Moreover, the extracellular loops, the C-terminal and the N-terminal also do not seem to be important in ligand recognition [60, 61].

Several series of agonists and antagonists for adenosine receptors have been developed by medicinal chemistry artwork either by structure-activity-relationship (SAR) or quantitative-structure-activity-relationship (QSAR) studies, which allowed the definition of structural and electrostatic requirements for the differential affinity of adenosine receptor subtypes binding. A brief summary of the pharmacological tools currently used for classification of adenosine receptors is presented in table I.

Table I. Pharmacological tools currently used for classification of adenosine receptors

Ligand	Abbreviation	Chemical name	Adenosine receptor subtype
AGONIST	ADO	Adenosine	A_1, A_{2A}, A_{2B}, A_3
	NECA	5'-*N*-ethyl-carboxamidoadenosine	A_1, A_{2A}, A_{2B}, A_3
	CPA	N^6-cyclopentyladenosine	A_1
	CCPA	2-chloro-CPA	A_1
	CGS 21680	2-p-(2-carboxyethyl)phenethylamino-5´-N-ethylcarboxamidoadenosine hydrochloride	A_{2A}
	IB-MECA	1-deoxy-1-[6-[[(3-iodophenyl)methyl]amino]-9H-purin-9-yl]-N-methyl-β-D-ribofuranuronamide	A_3
	2Cl-IB-MECA	2-chloro-N^6-(3-iodobenzyl)-5'-(N-methylcarbamoyl)adenosine	A_3
ANTAGONIST	Teophylline	3,7-Dihydro-1,3-dimethyl-1H-purine-2,6-dione	A_1, A_{2A}, A_{2B}, A_3
	DPCPX	1,3-dipropyl-8-cyclopentylxanthine	A_1
	SCH 58261	5-amino-7-(2-phenylethyl)-2-(2-furyl)-pyrazolo-[4,3-e]-1,2,4-triazolo[1,5-c]pyrimidine	A_{2A}
	ZM 241385	4-(2-[7-amino-2-[2-furyl]-[1,2,4]triazolo[2,3-*a*][1,3,5]triazin-5-yl-amino]ethyl)phenol	A_{2A}
	MRS 1754	*N*-(4-cyano-phenyl)-2-[4-(2,6-dioxo-1,3-dipropyl-2,3,4,5,6,7-hexahydro-1*H*-purin-8-yl)-phenoxy]acetamide	A_{2B}
	MRS 1220	N-[9-Chloro-2-(2-furanyl)[1,2,4]-triazolo[1,5-c]quinazolin-5-yl]benzene acetamide	A_3
	MRS 1523	2,3-diethyl-4,5-dipropyl-6-phenylpyridine-3-thiocarboxylate-5-carboxylate	A_3

3.3. Signaling Pathways

Adenosine receptors are metabotropic receptors that couple to heterotrimeric guanine nucleotide binding proteins (G-proteins). The intracellular segment of all adenosine receptor subtypes interacts with the appropriate G-protein with subsequent activation of the intracellular signal transduction mechanism. Currently, it is accepted that adenosine receptor subtypes, A_1, A_{2A}, A_{2B} and A_3, coupled to Gi/o, Gs, Gs/Gq and Gi/o/Gq, respectively [72]. Adenosine receptors are broadly grouped into two categories: adenosine A_1 and A_3 receptors, which couple to inhibitory G proteins, and A_{2A} and A_{2B} receptors, which couple to stimulatory G proteins. The third intracellular loop of the adenosine A_{2A} receptor seems to be the main determinant of its G-protein selectivity [73]. Nevertheless, it has been demonstrated that adenosine receptors are pleiotropic, i.e., they can couple with various G proteins and transduction systems according to their degree of activation and their particular cellular or subcellular localization [74].

3.4. Adenosine Receptors Distribution

The knowledge of receptors localization, distribution and function has been advanced through continuous improvement of methodologies and equipments. In this sense, the first approach developed to establish receptors localization was based on the use of radioligands and their binding visualized by *in vitro* autoradiography in tissue sections [75-77]. For instance, the work of Bruns and co-workers [78] was pioneer in identifying the localization of adenosine receptors. However, limitations in the accuracy of this method, at the cellular level, lead to the refinement of techniques such as RT-PCR or Western-blot and classic/fluorescent immunohistochemistry microscope visualization.

Notwithstanding the importance of qualitative information, reliable quantitative data related with the expression of receptors (or other type of proteins) can be precious and open possibilities, combined with functional data, to new perspectives of study, contributing for an integrative view of receptors/proteins, both in physiology and pharmacology. Currently, several methodologies, including laser scanning confocal microscopy (LSCM) have been used to characterize receptor dynamics: immunohistochemistry and analysis of signal transduction pathways, including visualization of protein binding to other molecules or their folding by using fluorescence resonance energy transfer (FRET) or bioluminescence resonance energy transfer (BRET) [79], live-cell imaging, including molecular dynamics by fluorescent protein-fusion probes [80, 81], traffic analysis of individual proteins, such as receptors, by fluorescence recovery after photobleaching (FRAP) techniques [82] and protein turnover.

In Table II, identification and (in some cases) characterization of the pattern of distribution of individual adenosine receptors in several tissues is depicted. The presence of adenosine receptor subtypes in tissues has also been documented for more than a decade: adenosine receptors were observed in vascular beds, both in arteries, such as the coronary artery [1, 83-85], pulmonary artery [86-89], mesenteric artery [89-91], tail artery [92-95], renal vasculature [89, 96, 97] and aorta [98-100], and veins [101-103].

Table II. Expression of adenosine receptors in tissues from several organs and systems

Expression	A_1 subtype	A_{2A} subtype	A_{2B} subtype	A_3 subtype
High	Atria, adrenal gland, spinal cord, brain [106]	Leukocytes, platelets, spleen, thymus, [107]	Bladder, Cecum, colon, [108]	Testis, mast cells [106]
Intermediate	Skeletal muscle, kidney, liver, adipocytes brain, [106]	Arteries and veins, heart, sympathetic/ parassympathetic nerves [107]	Lung, arteries and veins, mast cells [108]	Brain [106]
Low	Pancreas [106]	Brain [107]	Adipocytes, adrenal gland, brain, kidney [108]	Heart, thyroid, spleen, liver, kidney [106]

3.4.1. Vascular Wall: Intima, Media and Adventitia

The vascular wall layers can be identified by different morphologies and the different functions exhibited by respective cells. In this sense, the use of fluorescent nuclear stain will enable identification of cell orientation, number, organization and cell type allowing layer vessel recognition [104, 105]. Receptors can be located in any of these cells and may interplay with other receptors, enzymes or transporters located in the same cell or in neighbouring cells.

Both arteries and veins have a similar structure: three layers can be identified (intima, media and adventitia). Surrounding the lumen is the innermost layer, the tunica intima, which is the thinnest layer, made up of simple endothelial cells and lines the entire circulatory system. Also included in this thin layer are circular, elastic bands, the internal elastic lamina. The second layer, the tunica media, is made up of more circular elastic bands, the external elastic lamina. The tunica media contains vascular smooth muscle which helps regulate the size of the lumen and those of arteries contains more smooth muscle than the tunica media of their counterpart, the veins, and this allows arteries to constrict and dilate to adjust the volume of blood needed by the tissues that they support. Adventitia layer is composed of connective tissue that allows the blood vessel to withstand forces acting on the vessel wall. It also contains strong collagen fibers that help anchor the blood vessel to surrounding tissues, and this gives the vessel some stability.

Veins and arteries contain smooth muscle that is regulated by the sympathetic nervous system. In veins, nerves are distributed throughout the medial smooth muscle coat, often in close apposition (50 nm) to smooth muscle cells. In arteries, nerves are located at the adventitial-medial border, few closer than 2 nm to smooth muscle cells, often with interposing connective tissue and Schwann cell processes [109]. When necessary, the sympathetic nervous system signals arteries to relax or contract, which changes the lumen size and, thereby, regulates blood flow and also affects blood pressure. Sympathetic nerve fibers are enveloped in Schwann cells: most nerve fibers travel through individual channels in the Schwann cell, but small fibers are sometimes bundled together within a single channel [110].

3.4.2. Adenosine Receptors in the Vascular Wall

Over all the current knowledge of the function/localization of adenosine receptors or other molecular effectors in each vascular wall layer is based on the combination of data achieved from different methodological approaches.

Adenosine receptors located in cells from intima, media or adventitia can be identified using specific experimental approaches. For instance, to study adenosine receptors located in smooth muscle cells (media), measurement of vascular responsiveness (vasodilation/vasoconstriction) can be carried out. Another example is the study of adenosine receptors located in sympathetic fibers, in the adventitia, which can be done performing transmitter release experiments. Many other approaches can be designed depending on the specific goal of the researcher but, in some cases, removal of one or two of the vascular wall layers can be very useful. In this sense, many studies described endothelium removal by mechanical or chemical approaches [111, 112]. Adventitia removal has also been described to be experimentally possible with reliable results [113]. These latter techniques amplify the capacity of identifying and understanding the function of adenosine receptors or other proteins/signal mediators in cells located in each vascular wall layer.

Other type of insights is the interplay between adenosine receptors and signalling molecules such as reactive oxygen species (ROS), nitric oxide (NO), etc. Indeed, data indicates that adenosine receptor (A_{2A} subtype) activation promote the increase of ROS generation [114] having a role in oxidative stress and, consequently, in a large number of pathologies where oxidative stress is crucial. Moreover, adenosine receptors (adenosine A_1 aor A_{2A} receptors) may also activate eNOS leading to an increase of NO production [115], which may impair the deleterius effects mediated by ROS and oxidative stress.

3.4.2.1. Intima Layer

In endothelium adenosine receptor subtype A_1, A_{2A}, A_{2B} and A_3 have been identified by functional and immunohistochemical assays. Immunohistochemical studies allowed the identification of all adenosine receptor subtypes in the endothelium of several arteries such as tail artery [92, 98], and aorta [98, 116]. Nevertheless, few studies have characterised the presence of adenosine receptors in endothelium veins. In endothelial cells from Human Umbilical Vein (HUVEC) culture adenosine A_{2A} and A_{2B} receptors were related with cell proliferation in eclampsia and with thrombin-mediated proinflammatory signalling responses [117, 118] and their presence was confirmed by immunohistochemistry [119]. Adenosine, as a potent vasodilator in most of the vascular beds sustains the effect of Gestational Diabetes Mellitus (GDM) by activating adenosine receptors (likely adenosine A_{2B} receptors) [120]. Moreover, an increased plasma adenosine concentrations in the fetal blood from GDM pregnancies was detected. Adenosine is a modulator of L-arginine/NO pathway in these vessels and its effect most likely result from activation of plasma membrane receptors at the umbilical vein endothelium [121]. Nevertheless, more studies are needed to identify and better characterise the role of endothelium adenosine receptor subtypes in veins.

It is well established that endothelium may influence vascular responsiveness by producing vasoactive substances such as NO, ROS, endothelins and adenosine. In endothelium, it has been described that adenosine induces NO production through adenosine A_1 and A_{2A} receptor activation pathway which begins with activation of K^+_{ATP} channels that leads to an increase in $[Ca^{2+}]_i$, important in stimulating the phospholipase A_2, leading to liberation of arachidonic acid and stimulation of cycloogygenase pathway to generate prostaglandin I_2. The prostaglandin I_2 is then released from cell and acts on its endothelial prostacyclin receptors which are linked to adenylyl cyclase to increase intracellular cAMP, activate protein kinase A and so phosphorylate and activate endothelial nitric oxide synthase [115]. Therefore, adenosine stimulate endothelial NO synthase activity, which in turn, generates higher amounts of NO, a well-known vasodilator [122, 123]. Endothelium-derived adenosine was also described to activate prejunctional adenosine receptors, mainly A_1 and A_{2A}, which modulate neurotransmission influencing vascular tonus [124].

Adenosine A_{2B} receptor subtype, in endothelial cells, was implicated in cell proliferation [125], suggesting that pharmacological or molecular activation of this receptor subtype may be useful in modulating vascular remodelling. Adenosine A_{2B} receptor is, therefore, a protective effector against hyperplasia. Moreover, adenosine A_{2B} receptors were found to be highly expressed in macrophages and vascular smooth muscle cells presenting an important role in the regulation of inflammation and vascular adhesion: deficiency in adenosine A_{2B} receptors was shown to promote lesions or thickness of the neointima after vascular injury [126] revealing its protective role in atherosclerosis.

Some studies also identified adenosine A_{2A} [118] and A_3 [127] receptors as being protective against endothelial injury induced by inflammatory processes.

3.4.2.2. Media Layer

In vascular tissues adenosine is known to induce vasodilatation and this effect is classically ascribed to A_2 receptor-mediated effects on vascular smooth muscle cells, thus increasing blood flow and oxygenation; also, adenosine released during preconditioning by short periods of ischemia followed by reperfusion can induce cardioprotection to subsequent sustained ischemia [128]. In fact adenosine evokes dilatation of vascular smooth muscle that occur by stimulating adenosine A_2 receptor subtypes, mainly the A_{2A} subtype (linked to K_{ATP} sensitive channels), an effect caused by an increase in cAMP [129]. Nevertheless, evidence has also demonstrated that adenosine A_1 receptor subtype can also mediate vasodilatation [130-133] in an endothelium- and nitric oxide-dependent [134] fashion. Indeed, in more recent studies, it was demonstrated that A_1 and A_{2A} receptor activation in endothelial cells promotes NO production and, consequently, NO-mediated vasodilation.

Adenosine A_{2B} receptors have been described to cause vasodilation in vascular smooth muscle cells from vessels such as aorta and saphenous vein.

(135) On the other hand, no evidence has been linked to A_3 receptor activation producing relaxation of blood vessels.

Accordingly with functional studies, immunohistochemistry have demonstrated the presence of all adenosine receptor subtypes in smooth muscle cells of arterial media layer [98]. In veins, few studies related with adenosine receptors are available and, to our knowledge, until to date, the distribution profile of adenosine receptor subtypes in smooth muscle cells from veins is still to be done. A broad number of studies reported antimitogenic effects to adenosine, *via* activation of adenosine A_{2B} receptors in pulmonary [86], aorta [136] and glomerular [137] artery smooth muscle cells. Vascular smooth muscle proliferation can be inhibited through cAMP/ Epac (exchange protein directly activated by cAMP) [138] pathway after adenosine A_{2B} receptor activation which, in turn, can be regulated by B-Myb transcription factor [139]. More recently and, by contrast, adenosine A_1 receptor was found to promote coronary smooth muscle cells proliferation [140]. These opposite effects ascribed to adenosine in the media layer of arteries, mediated by different adenosine receptor subtypes, evidence the importance of adenosine levels, a crucial factor determining the protective or promoter role of adenosine. Therefore, depending on the subtype of adenosine receptor activated, inhibition or stimulation of smooth muscle cells hypertrophy may occur.

3.4.2.3. Adventitial Layer

Several studies described the occurrence of a neuromodulatory role ascribed to adenosine receptor subtypes activation in sympathetic nerve fibers located in the adventitia layer of pulmonary [141], mesenteric [90, 91, 101, 103, 142], ear [142], aorta [143], tail [92-95, 144] arteries as well as in veins such as mesenteric veins [101, 102]. Accordingly, in renal vasculature the influence of adenosine receptors in sympathetic regulation was also described [145] conditioning the blood efflux in afferent arteriole, and therefore, renal filtration.

From immunohistochemical studies it was possible to identify the presence of adenosine A_1 (tail artery [98], mesenteric artery and vein [90, 91, 102]), A_{2A} and A_{2B} (tail and mesenteric artery and vein [91, 92, 98]) and A_3 (mesenteric vein [101]) receptors. Recent studies allowed the visualization of adenosine receptor subtypes (A_1, A_{2A}, A_{2B}, A_3) in sympathetic nerve fibers. [90, 91, 102] Taken together, correlated morphological and functional data allowed advances into the insights of neurovascular sympathetic modulation mediated by adenosine receptors, particularly in pathological conditions such as hypertension.

In hypertensive arteries and veins there are an impairment in the neuromodulatory effects mediated by adenosine A_1 receptors [90, 91, 101, 102], while the adenosine A_{2A} receptor-mediated facilitation of noradrenaline release is preserved [90, 91, 101]. Moreover, a redistribution of adenosine A_1 receptors from sympathetic nerves to Schwann cells was observed in hypertensive state supporting the functional data. Concerning the amounts of adenosine A_{2A} receptors present in sympathetic nerves, they were preserved in hypertension.

The idea that endothelium could influence neurotransmission [146] was recently supported by findings where endogenous adenosine (derived from endothelium) altered neurotransmission (mesenteric and tail arteries).

Adenosine A_{2B} and A_3 receptors in vessels seem to have an important role in conditions where amounts of adenosine are higher, i.e, in pathological conditions such as in hypertension [147, 148], diabetes [149-151]. Adenosine A_{2B} receptors increase noradrenaline release in vascular tissues such as tail arteries [92]. Adenosine A_3 receptors have the opposite effect, inhibiting the release of neurotransmitters [152-154].

Hypertrophy is a consequence of the complex interaction of intravascular pressure, systemic and local hormones, genetic predispositions, and environmental factors. Animal and human studies demonstrate that sustained increases in sympathetic activation directly induce vascular remodeling. Therefore, adenosine A_2 receptors, known to facilitate noradrenaline release, may have a profound impact in vascular remodeling, by enhancing noradrenaline levels in the synaptic cleft. Many studies report greater circulating levels of noradrenaline in patients with hypertension than in normotensive control subjects. In normotensive subjects, increased levels of circulating noradrenaline generally induce down-regulation of noradrenergic receptors. However, in subjects with hypertension, such down-regulation appears not to occur, resulting in enhanced sensitivity to noradrenaline. The combination of enhanced sensitivity to and increased circulating levels of noradrenaline likely contributes significantly to sympathetic nervous system activity-related hypertension.

An important effect mediate by adenosine in adventitia is also related with the role of adenosine receptors in inflammation. Indeed, adventitia present several cells involved in inflammatory processes: macrophages, lymphocytes, fibrocytes. Such cells were found to present adenosine receptor subtypes: neutrophils, eosinophils, and macrophages exhibited adenosine A_3 receptors revealing anti-inflammatory effects, while in basophils this adenosine receptor subtype was found to be pro-inflammatory [155]; in macrophages and lymphocytes adenosine A_2 receptors were also described to be expressed. [156, 157].

4. Therapeutic Potential of Adenosine Receptor Ligands

Until recently, most, if not all, pharmacological studies have been based on the development of agonist or antagonists and the targeting of the signalling cascades triggered by membrane receptor modulation. After years of basic and clinical research it became evident that adenosine may have a potential therapeutic utility is several pathological conditions (Table III). At present research efforts developed both in medicinal chemistry and pharmacology/pathophysiology fields led to several new adenosine receptor ligands, agonists and antagonists, which are under clinical evaluation for different therapeutic applications.

Studies with capadenoson (phase II), a new adenosine A_1 receptor agonist, showed its utility to treat atrial fibrillation. [158] Tecadenoson, is another recent adenosine A_1 receptor agonist (phase II) evaluated for its therapeutic activity in converting paroxysmal supraventricular tachycardia (PSVT) to sinus rhythm. [159] Tecadenoson can be transported by hENT1 and therefore, it could be used to target central nervous system disorders because of its enhanced delivery to the brain. [160]

New agonists and antagonists for adenosine A_{2A} receptors have also been developed: the adenosine A_{2A} agonists, apadenoson (Phase III), binodenoson (phase II) and regadenoson are used in myocardial perfusion imaging [161], and MRE0094, as promoter of diabetic foot ulcers healing. [162] Istradefylline [163] (phase III) and prelandenant [164] (Phase III) are new selective A_{2A} receptor antagonists, recently approved for the treatment of Parkinson´s disease.

Adenosine A_{2B} receptor antagonists have been tested as a useful pharmacological tool in inflammatory diseases [177], namely in asthma, chronic obstructive pulmonary disease, and pulmonary fibrosis (CVT- 6883: [178, 179]). Concerning new ligands for adenosine A_3 receptors, CF101 (IB-MECA in Phase II/III) [180, 181] and CF102 (Cl-IB-MECA in Phase I/II)

[182] are currently under evaluation for inflammatory disorders and cancer, respectively.

Table III. Therapeutic potential of adenosine receptor ligands

		Therapeutic potential
Adenosine receptor agonists	A_1	Atrioventricular node block and supraventricular tachyarrhythmia [165]
	A_{2A}	Respiratory disorders, sepsis, reperfusion injury, thrombosis, hypertension, inflammatory disorders [166-170]
	A_{2B}	Allergic reactions [171]
	A_3	Cardiac ischemias, arrhythmias [167, 169]
Adenosine receptor antagonists	A_1	Bradyarrhythmia associated with inferior myocardial infarction, cardiac arrest, cardiac transplant rejection [172]
	A_{2A}	Parkinson's disease, neurodegeneration [166, 173]
	A_{2B}	Asthma, pulmonary inflammation [174]
	A_3	Glaucoma, asthma [174-176]

Some of these promising candidates might be discarded as so due to side effects. For example, concerning regadenoson, Food and Drug Administration (FDA) recently recommended that physicians should avoid its use for cardiac nuclear stress tests in patients with signs/symptoms of unstable angina or cardiovascular instability because this drug may increase the risk for a fatal heart attack. Nevertheless, and based on growing scientific evidence, some of these ligands will be approved for clinical use and it is expected that they will significantly alter the life styles and outcomes of patients.

Conclusion

Several studies have raised the possibility that receptors might be located in cellular compartments other than the plasma membrane. Adenosine receptors should be viewed as dynamic proteins in constant traffic within the plasma membrane or among other organelle membranes where they can be stimulated by exogenous or endogenous ligands. Receptor activity is modulated by fine tunning and their steady state is balanced by a mechanism involving receptor internalization, through endocytosis. Then, receptors are routed to endosomes and they can be either recycled into its "active state" into membranes for repeated receptor activation [183]) or be degradated inside lysosomes, which lead to the termination of receptor signalling mediated events [184-187].

This perspective can increase the pharmacological understanding of drugs and lead to the development of new therapeutic approaches, particularly if their physico-chemical properties enable them to cross the plasma membrane. Indeed, in the last 15 years several authors have investigated the ocurrence of several types of receptors located inside the cell [194-196], including adenosine receptors [188, 189], both in cytoplasmatic organelles and in the nuclei, helping in the understanding of the subsequent signalling events triggered by their activation.

References

[1] Headrick JP, Ashton KJ, Rose'meyer RB, Peart JN. Cardiovascular adenosine receptors: expression, actions and interactions. *Pharmacology & therapeutics*. 2013;140(1):92-111.

[2] Conlon BA, Ross JD, Law WR. Advances in understanding adenosine as a plurisystem modulator in sepsis and the systemic inflammatory response syndrome (SIRS). *Front Biosci*. 2005;10:2548-65.

[3] Meghji P, Pearson JD, Slakey LL. Regulation of extracellular adenosine production by ectonucleotidases of adult rat ventricular myocytes. *Am. J. Physiol.* 1992;263(1 Pt 2):H40-7.

[4] Fredholm BB. Adenosine receptors as targets for drug development. *Drug News Perspect*. 2003;16(5):283-9.

[5] Endres CJ, Moss AM, Govindarajan R, Choi DS, Unadkat JD. The role of nucleoside transporters in the erythrocyte disposition and oral absorption of ribavirin in the wild-type and equilibrative nucleoside transporter 1-/- mice. *J. Pharmacol. Exp. Ther*. 2009;331(1):287-96.

[6] Molina-Arcas M, Casado FJ, Pastor-Anglada M. Nucleoside transporter proteins. *Curr. Vasc. Pharmacol*. 2009;7(4):426-34.

[7] Molina-Arcas M, Trigueros-Motos L, Casado FJ, Pastor-Anglada M. Physiological and pharmacological roles of nucleoside transporter proteins. *Nucleosides Nucleotides Nucleic Acids*. 2008;27(6):769-78.

[8] Baldwin SA, Beal PR, Yao SY, King AE, Cass CE, Young JD. The equilibrative nucleoside transporter family, SLC29. *Pflugers Arch.* 2004;447(5):735-43.

[9] Lu H, Chen C, Klaassen C. Tissue distribution of concentrative and equilibrative nucleoside transporters in male and female rats and mice. *Drug metabolism and disposition: the biological fate of chemicals.* 2004;32(12):1455-61.

[10] Anderson CM, Xiong W, Geiger JD, Young JD, Cass CE, Baldwin SA, et al. Distribution of equilibrative, nitrobenzylthioinosine-sensitive nucleoside transporters (ENT1) in brain. *J. Neurochem.* 1999;73(2):867-73.

[11] Cass CE, Young JD, Baldwin SA, Cabrita MA, Graham KA, Griffiths M, et al. Nucleoside transporters of mammalian cells. *Pharm. Biotechnol.* 1999;12:313-52.

[12] Huber-Ruano I, Pastor-Anglada M. Transport of nucleoside analogs across the plasma membrane: a clue to understanding drug-induced cytotoxicity. *Curr. Drug Metab.* 2009;10(4):347-58.

[13] Molina-Arcas M, Marce S, Villamor N, Huber-Ruano I, Casado FJ, Bellosillo B, et al. Equilibrative nucleoside transporter-2 (hENT2) protein expression correlates with ex vivo sensitivity to fludarabine in chronic lymphocytic leukemia (CLL) cells. *Leukemia.* 2005;19(1):64-8.

[14] Yao SY, Ng AM, Vickers MF, Sundaram M, Cass CE, Baldwin SA, et al. Functional and molecular characterization of nucleobase transport by recombinant human and rat equilibrative nucleoside transporters 1 and 2. Chimeric constructs reveal a role for the ENT2 helix 5-6 region in nucleobase translocation. *J. Biol. Chem.* 2002;277(28):24938-48.

[15] Podgorska M, Kocbuch K, Pawelczyk T. Recent advances in studies on biochemical and structural properties of equilibrative and concentrative nucleoside transporters. *Acta Biochim. Pol.* 2005;52(4):749-58.

[16] Young JD, Yao SY, Baldwin JM, Cass CE, Baldwin SA. The human concentrative and equilibrative nucleoside transporter families, SLC28 and SLC29. *Molecular aspects of medicine.* 2013;34(2-3):529-47.

[17] Hyde RJ, Cass CE, Young JD, Baldwin SA. The ENT family of eukaryote nucleoside and nucleobase transporters: recent advances in the investigation of structure/function relationships and the identification of novel isoforms. *Mol. Membr Biol.* 2001;18(1):53-63.

[18] Cass CE, Young JD, Baldwin SA. Recent advances in the molecular biology of nucleoside transporters of mammalian cells. *Biochem. Cell Biol.* 1998;76(5):761-70.

[19] Kong W, Engel K, Wang J. Mammalian nucleoside transporters. *Curr. Drug Metab.* 2004;5(1):63-84.

[20] Barnes K, Dobrzynski H, Foppolo S, Beal PR, Ismat F, Scullion ER, et al. Distribution and functional characterization of equilibrative nucleoside transporter-4, a novel cardiac adenosine transporter activated at acidic pH. *Circulation research*. 2006;99(5):510-9.

[21] Ritzel MW, Ng AM, Yao SY, Graham K, Loewen SK, Smith KM, et al. Recent molecular advances in studies of the concentrative Na+-dependent nucleoside transporter (CNT) family: identification and characterization of novel human and mouse proteins (hCNT3 and mCNT3) broadly selective for purine and pyrimidine nucleosides (system cib). *Mol. Membr Biol*. 2001;18(1):65-72.

[22] Errasti-Murugarren E, Molina-Arcas M, Casado FJ, Pastor-Anglada M. A splice variant of the SLC28A3 gene encodes a novel human concentrative nucleoside transporter-3 (hCNT3) protein localized in the endoplasmic reticulum. *FASEB journal : official publication of the Federation of American Societies for Experimental Biology*. 2009;23(1):172-82.

[23] Gloeckner-Hofmann K, Guillen-Gomez E, Schmidtgen C, Porstmann R, Ziegler R, Stoss O, et al. Expression of the high-affinity fluoropyrimidine-preferring nucleoside transporter hCNT1 correlates with decreased disease-free survival in breast cancer. *Oncology*. 2006;70(3):238-44.

[24] Peng XX, Shi Z, Damaraju VL, Huang XC, Kruh GD, Wu HC, et al. Up-regulation of MRP4 and down-regulation of influx transporters in human leukemic cells with acquired resistance to 6-mercaptopurine. *Leuk Res*. 2008;32(5):799-809.

[25] Ritzel MW, Yao SY, Huang MY, Elliott JF, Cass CE, Young JD. Molecular cloning and functional expression of cDNAs encoding a human Na+-nucleoside cotransporter (hCNT1). *Am. J. Physiol.* 1997;272(2 Pt 1):C707-14.

[26] Ritzel MW, Yao SY, Ng AM, Mackey JR, Cass CE, Young JD. Molecular cloning, functional expression and chromosomal localization of a cDNA encoding a human Na+/nucleoside cotransporter (hCNT2) selective for purine nucleosides and uridine. *Mol. Membr Biol.* 1998;15(4):203-11.

[27] Ward JL, Sherali A, Mo ZP, Tse CM. Kinetic and pharmacological properties of cloned human equilibrative nucleoside transporters, ENT1 and ENT2, stably expressed in nucleoside transporter-deficient PK15 cells. Ent2 exhibits a low affinity for guanosine and cytidine but a high affinity for inosine. *J. Biol. Chem*. 2000;275(12):8375-81.

[28] Yao SY, Ng AM, Muzyka WR, Griffiths M, Cass CE, Baldwin SA, et al. Molecular cloning and functional characterization of nitrobenzylthioinosine (NBMPR)-sensitive (es) and NBMPR-insensitive (ei) equilibrative nucleoside transporter proteins (rENT1 and rENT2) from rat tissues. *J. Biol. Chem.* 1997;272(45):28423-30.
[29] Parkinson FE, Damaraju VL, Graham K, Yao SY, Baldwin SA, Cass CE, et al. Molecular biology of nucleoside transporters and their distributions and functions in the brain. *Curr. Top Med. Chem.* 2011;11(8):948-72.
[30] King AE, Ackley MA, Cass CE, Young JD, Baldwin SA. Nucleoside transporters: from scavengers to novel therapeutic targets. *Trends in pharmacological sciences*. 2006;27(8):416-25.
[31] Zhang J, Visser F, King KM, Baldwin SA, Young JD, Cass CE. The role of nucleoside transporters in cancer chemotherapy with nucleoside drugs. *Cancer metastasis reviews*. 2007;26(1):85-110.
[32] Damaraju VL, Smith KM, Mowles D, Nowak I, Karpinski E, Young JD, et al. Interaction of fused-pyrimidine nucleoside analogs with human concentrative nucleoside transporters: High-affinity inhibitors of human concentrative nucleoside transporter 1. *Biochem. Pharmacol.* 2011;81(1):82-90.
[33] Drury AN, Szent-Gyorgyi A. The physiological activity of adenine compounds with especial reference to their action upon the mammalian heart. *J. Physiol.* 1929;68(3):213-37.
[34] Sebastiao AM, Ribeiro JA. Fine-tuning neuromodulation by adenosine. *Trends in pharmacological sciences*. 2000;21(9):341-6.
[35] Burnstock G. Cotransmission. *Current opinion in pharmacology.* 2004;4(1):47-52.
[36] Diniz C, Borges F, Santana L, Uriarte E, Oliveira JM, Goncalves J, et al. Ligands and therapeutic perspectives of adenosine A(2A) receptors. *Current pharmaceutical design*. 2008;14(17):1698-722.
[37] Londos C, Wolff J. Two distinct adenosine-sensitive sites on adenylate cyclase. *Proc. Natl. Acad. Sci. USA*. 1977;74(12):5482-6.
[38] van Calker D, Muller M, Hamprecht B. Adenosine inhibits the accumulation of cyclic AMP in cultured brain cells. *Nature.* 1978;276(5690):839-41.
[39] Londos C, Cooper DM, Wolff J. Subclasses of external adenosine receptors. *Proc. Natl. Acad. Sci. USA*. 1980;77(5):2551-4.

[40] Fredholm BB, Abbracchio MP, Burnstock G, Daly JW, Harden TK, Jacobson KA, et al. Nomenclature and classification of purinoceptors. *Pharmacol. Rev.* 1994;46(2):143-56.

[41] Daly JW, Butts-Lamb P, Padgett W. Subclasses of adenosine receptors in the central nervous system: interaction with caffeine and related methylxanthines. *Cell Mol. Neurobiol.* 1983;3(1):69-80.

[42] Libert F, Passage E, Parmentier M, Simons MJ, Vassart G, Mattei MG. Chromosomal mapping of A1 and A2 adenosine receptors, VIP receptor, and a new subtype of serotonin receptor. *Genomics.* 1991;11(1):225-7.

[43] Maenhaut C, Van Sande J, Libert F, Abramowicz M, Parmentier M, Vanderhaegen JJ, et al. RDC8 codes for an adenosine A2 receptor with physiological constitutive activity. *Biochemical and biophysical research communications.* 1990;173(3):1169-78.

[44] Stehle JH, Rivkees SA, Lee JJ, Weaver DR, Deeds JD, Reppert SM. Molecular cloning and expression of the cDNA for a novel A2-adenosine receptor subtype. *Molecular endocrinology.* 1992;6(3):384-93.

[45] Jacobson MA, Johnson RG, Luneau CJ, Salvatore CA. Cloning and chromosomal localization of the human A2b adenosine receptor gene (ADORA2B) and its pseudogene. *Genomics.* 1995;27(2):374-6.

[46] Meyerhof W, Paust HJ, Schonrock C, Richter D. Cloning of a cDNA encoding a novel putative G-protein-coupled receptor expressed in specific rat brain regions. *DNA Cell Biol.* 1991;10(9):689-94.

[47] Zhou QY, Li C, Olah ME, Johnson RA, Stiles GL, Civelli O. Molecular cloning and characterization of an adenosine receptor: the A3 adenosine receptor. *Proc. Natl. Acad. Sci. USA.* 1992;89(16):7432-6.

[48] Klotz K-N. Pharmacology and Molecular Biology of A3 Adenosine Receptors. In: Borea PA, editor. A3 Adenosine Receptors from Cell Biology to Pharmacology and Therapeutics: Springer Dordrecht Heidelberg London New York; 2010.

[49] Dunwiddie TV, Masino SA. The role and regulation of adenosine in the central nervous system. *Annu. Rev. Neurosci.* 2001;24:31-55.

[50] Cieslak M, Komoszynski M, Wojtczak A. Adenosine A(2A) receptors in Parkinson's disease treatment. *Purinergic Signal.* 2008;4(4):305-12.

[51] Szkotak AJ, Ng AM, Man SF, Baldwin SA, Cass CE, Young JD, et al. Coupling of CFTR-mediated anion secretion to nucleoside transporters and adenosine homeostasis in Calu-3 cells. *The Journal of membrane biology.* 2003;192(3):169-79.

[52] Choi DS, Cascini MG, Mailliard W, Young H, Paredes P, McMahon T, et al. The type 1 equilibrative nucleoside transporter regulates ethanol intoxication and preference. *Nature neuroscience*. 2004;7(8):855-61.
[53] Chen J, Nam HW, Lee MR, Hinton DJ, Choi S, Kim T, et al. Altered glutamatergic neurotransmission in the striatum regulates ethanol sensitivity and intake in mice lacking ENT1. *Behav. Brain Res.* 2010;208(2):636-42.
[54] Nam HW, Lee MR, Zhu Y, Wu J, Hinton DJ, Choi S, et al. Type 1 equilibrative nucleoside transporter regulates ethanol drinking through accumbal N-methyl-D-aspartate receptor signaling. *Biological psychiatry*. 2011;69(11):1043-51.
[55] Chen J, Rinaldo L, Lim SJ, Young H, Messing RO, Choi DS. The type 1 equilibrative nucleoside transporter regulates anxiety-like behavior in mice. *Genes, brain, and behavior*. 2007;6(8):776-83.
[56] Kost S, Sun C, Xiong W, Graham K, Cass CE, Young JD, et al. Behavioral effects of elevated expression of human equilibrative nucleoside transporter 1 in mice. *Behav. Brain Res.* 2011;224(1):44-9.
[57] Rose JB, Naydenova Z, Bang A, Eguchi M, Sweeney G, Choi DS, et al. Equilibrative nucleoside transporter 1 plays an essential role in cardioprotection. *American journal of physiology Heart and circulatory physiology*. 2010;298(3):H771-7.
[58] Sousa JB FP, Diniz C. Endothelial dysfunction impairs vascular neurotransmission in tail arteries. *Neurochemistry International*. 2014.
[59] Ralevic V, Burnstock G. Receptors for purines and pyrimidines. *Pharmacol Rev*. 1998;50(3):413-92.
[60] Olah ME, Jacobson KA, Stiles GL. Role of the second extracellular loop of adenosine receptors in agonist and antagonist binding. Analysis of chimeric A1/A3 adenosine receptors. *J. Biol. Chem.* 1994;269(40):24692-8.
[61] Olah ME, Ren H, Stiles GL. Adenosine receptors: protein and gene structure. *Archives internationales de pharmacodynamie et de therapie*. 1995;329(1):135-50.
[62] Olah ME, Jacobson KA, Stiles GL. Identification of an adenosine receptor domain specifically involved in binding of 5'-substituted adenosine agonists. *J. Biol. Chem.* 1994;269(27):18016-20.
[63] Olah ME, Ren H, Ostrowski J, Jacobson KA, Stiles GL. Cloning, expression, and characterization of the unique bovine A1 adenosine receptor. Studies on the ligand binding site by site-directed mutagenesis. *J. Biol. Chem.* 1992;267(15):10764-70.

[64] Kim J, Wess J, van Rhee AM, Schoneberg T, Jacobson KA. Site-directed mutagenesis identifies residues involved in ligand recognition in the human A2a adenosine receptor. *J. Biol. Chem.* 1995;270(23):13987-97.

[65] Gao ZG, Jiang Q, Jacobson KA, Ijzerman AP. Site-directed mutagenesis studies of human A(2A) adenosine receptors: involvement of glu(13) and his(278) in ligand binding and sodium modulation. *Biochem Pharmacol.* 2000;60(5):661-8.

[66] Townsend-Nicholson A, Schofield PR. A threonine residue in the seventh transmembrane domain of the human A1 adenosine receptor mediates specific agonist binding. *J. Biol. Chem.* 1994;269(4):2373-6.

[67] Beukers MW, den Dulk H, van Tilburg EW, Brouwer J, Ijzerman AP. Why are A(2B) receptors low-affinity adenosine receptors? Mutation of Asn273 to Tyr increases affinity of human A(2B) receptor for 2-(1-Hexynyl)adenosine. *Molecular pharmacology*. 2000;58(6):1349-56.

[68] Jiang Q, Van Rhee AM, Kim J, Yehle S, Wess J, Jacobson KA. Hydrophilic side chains in the third and seventh transmembrane helical domains of human A2A adenosine receptors are required for ligand recognition. *Molecular pharmacology*. 1996;50(3):512-21.

[69] Tucker AL, Robeva AS, Taylor HE, Holeton D, Bockner M, Lynch KR, et al. A1 adenosine receptors. Two amino acids are responsible for species differences in ligand recognition. *J. Biol. Chem.* 1994;269(45):27900-6.

[70] AP IJ, Von Frijtag Drabbe Kunzel JK, Kim J, Jiang Q, Jacobson KA. Site-directed mutagenesis of the human adenosine A2A receptor. Critical involvement of Glu13 in agonist recognition. *Eur. J. Pharmacol.* 1996;310(2-3):269-72.

[71] Scholl DJ, Wells JN. Serine and alanine mutagenesis of the nine native cysteine residues of the human A(1) adenosine receptor. *Biochem. Pharmacol.* 2000;60(11):1647-54.

[72] Fredholm BB, AP IJ, Jacobson KA, Klotz KN, Linden J. International Union of Pharmacology. XXV. Nomenclature and classification of adenosine receptors. *Pharmacol. Rev.* 2001;53(4):527-52.

[73] Olah ME. Identification of A2a adenosine receptor domains involved in selective coupling to Gs. Analysis of chimeric A1/A2a adenosine receptors. *J. Biol. Chem.* 1997;272(1):337-44.

[74] Cunha RA. Neuroprotection by adenosine in the brain: From A(1) receptor activation to A (2A) receptor blockade. *Purinergic Signal.* 2005;1(2):111-34.

[75] Rebola N, Sebastiao AM, de Mendonca A, Oliveira CR, Ribeiro JA, Cunha RA. Enhanced adenosine A2A receptor facilitation of synaptic transmission in the hippocampus of aged rats. *Journal of neurophysiology*. 2003;90(2):1295-303.

[76] Queiroz G, Talaia C, Goncalves J. Adenosine A2A receptor-mediated facilitation of noradrenaline release involves protein kinase C activation and attenuation of presynaptic inhibitory receptor-mediated effects in the rat vas deferens. *J Neurochem*. 2003;85(3):740-8.

[77] Lopes LV, Cunha RA, Kull B, Fredholm BB, Ribeiro JA. Adenosine A(2A) receptor facilitation of hippocampal synaptic transmission is dependent on tonic A(1) receptor inhibition. *Neuroscience*. 2002;112(2):319-29.

[78] Bruns RF, Lu GH, Pugsley TA. Characterization of the A2 adenosine receptor labeled by [3H]NECA in rat striatal membranes. *Molecular pharmacology*. 1986;29(4):331-46.

[79] Castillo-Badillo JA, Cabrera-Wrooman A, Garcia-Sainz JA. Visualizing G protein-coupled receptors in action through confocal microscopy techniques. *Archives of medical research*. 2014;45(4):283-93.

[80] Hori Y, Kikuchi K. Protein labeling with fluorogenic probes for no-wash live-cell imaging of proteins. *Current opinion in chemical biology*. 2013;17(4):644-50.

[81] Kobilka B, Gether U, Seifert R, Lin S, Ghanouni P. Characterization of ligand-induced conformational states in the beta 2 adrenergic receptor. *Journal of receptor and signal transduction research*. 1999;19(1-4):293-300.

[82] Calizo RC, Scarlata S. Discrepancy between fluorescence correlation spectroscopy and fluorescence recovery after photobleaching diffusion measurements of G-protein-coupled receptors. *Analytical biochemistry*. 2013;440(1):40-8.

[83] Teng B, Ledent C, Mustafa SJ. Up-regulation of A 2B adenosine receptor in A 2A adenosine receptor knockout mouse coronary artery. *Journal of molecular and cellular cardiology*. 2008;44(5):905-14.

[84] Ansari HR, Teng B, Nadeem A, Roush KP, Martin KH, Schnermann J, et al. A(1) adenosine receptor-mediated PKC and p42/p44 MAPK signaling in mouse coronary artery smooth muscle cells. *American journal of physiology Heart and circulatory physiology*. 2009;297(3):H1032-9.

[85] Hinze AV, Mayer P, Harst A, von Kugelgen I. Adenosine A(3) receptor-induced proliferation of primary human coronary smooth muscle cells

involving the induction of early growth response genes. *Journal of molecular and cellular cardiology*. 2012;53(5):639-45.

[86] Qian G, Cao J, Chen C, Wang L, Huang X, Ding C, et al. Paeoniflorin inhibits pulmonary artery smooth muscle cells proliferation via upregulating A2B adenosine receptor in rat. *PLoS One*. 2013;8(7):e69141.

[87] Umapathy SN, Kaczmarek E, Fatteh N, Burns N, Lucas R, Stenmark KR, et al. Adenosine A1 receptors promote vasa vasorum endothelial cell barrier integrity via Gi and Akt-dependent actin cytoskeleton remodeling. *PLoS One*. 2013;8(4):e59733.

[88] Alencar AK, Pereira SL, Montagnoli TL, Maia RC, Kummerle AE, Landgraf SS, et al. Beneficial effects of a novel agonist of the adenosine A2A receptor on monocrotaline-induced pulmonary hypertension in rats. *Br. J. Pharmacol*. 2013;169(5):953-62.

[89] Zhu Y, Liu L, Peng X, Ding X, Yang G, Li T. Role of adenosine A2A receptor in organ-specific vascular reactivity following hemorrhagic shock in rats. *J. Surg. Res*. 2013;184(2):951-8.

[90] Rocha-Pereira C, Arribas SM, Fresco P, Gonzalez MC, Goncalves J, Diniz C. Impaired inhibitory function of presynaptic A1-adenosine receptors in SHR mesenteric arteries. *Journal of pharmacological sciences*. 2013;122(2):59-70.

[91] Sousa JB, Vieira-Rocha MS, Sa C, Ferreirinha F, Correia-de-Sa P, Fresco P, et al. Lack of Endogenous Adenosine Tonus on Sympathetic Neurotransmission in Spontaneously Hypertensive Rat Mesenteric Artery. *PLoS One*. 2014;9(8):e105540.

[92] Diniz C, Fresco P, Leal S, Goncalves J. Adenosine receptors involved in modulation of noradrenaline release in isolated rat tail artery. *Eur. J. Pharmacol.* 2004;504(1-2):17-25.

[93] Fresco P, Diniz C, Goncalves J. Facilitation of noradrenaline release by activation of adenosine A(2A) receptors triggers both phospholipase C and adenylate cyclase pathways in rat tail artery. *Cardiovasc. Res.* 2004;63(4):739-46.

[94] Fresco P, Diniz C, Queiroz G, Goncalves J. Release inhibitory receptors activation favours the A2A-adenosine receptor-mediated facilitation of noradrenaline release in isolated rat tail artery. *Br. J. Pharmacol.* 2002;136(2):230-6.

[95] Fresco P, Oliveira JM, Kunc F, Soares AS, Rocha-Pereira C, Goncalves J, et al. A2A adenosine-receptor-mediated facilitation of noradrenaline release in rat tail artery involves protein kinase C activation and

betagamma subunits formed after alpha2-adrenoceptor activation. *Neurochem. Int.* 2007;51(1):47-56.

[96] Persson P, Hansell P, Palm F. Adenosine A2 receptor-mediated regulation of renal hemodynamics and glomerular filtration rate is abolished in diabetes. *Advances in experimental medicine and biology.* 2013;765:225-30.

[97] Booth LC, Tummers L, Jensen EC, Barrett CJ, Malpas SC, Gunn AJ, et al. Differential effects of the adenosine A1 receptor agonist adenosine amine congener on renal, femoral and carotid vascular conductance in preterm fetal sheep. *Clin. Exp. Pharmacol. Physiol.* 2008;35(11):1316-20.

[98] Leal S, Sa C, Goncalves J, Fresco P, Diniz C. Immunohistochemical characterization of adenosine receptors in rat aorta and tail arteries. *Microsc Res. Tech.* 2008;71(10):703-9.

[99] Teng B, Smith JD, Rosenfeld ME, Robinet P, Davis ME, Morrison RR, et al. A(1) adenosine receptor deficiency or inhibition reduces atherosclerotic lesions in apolipoprotein E deficient mice. *Cardiovasc. Res.* 2014;102(1):157-65.

[100] Pradhan I, Zeldin DC, Ledent C, Mustafa JS, Falck JR, Nayeem MA. High salt diet exacerbates vascular contraction in the absence of adenosine A(2)A receptor. *J. Cardiovasc. Pharmacol.* 2014;63(5):385-94.

[101] Rocha-Pereira C, Sousa JB, Vieira-Rocha MS, Fresco P, Goncalves J, Diniz C. Differential inhibition of noradrenaline release mediated by inhibitory A(1)-adenosine receptors in the mesenteric vein and artery from normotensive and hypertensive rats. *Neurochem. Int.* 2013;62(4):399-405.

[102] Sangsiri S, Dong H, Swain GM, Galligan JJ, Xu H. Impaired function of prejunctional adenosine A1 receptors expressed by perivascular sympathetic nerves in DOCA-salt hypertensive rats. *J. Pharmacol. Exp.* Ther. 2013;345(1):32-40.

[103] Talaia C, Morato M, Quintas C, Goncalves J, Queiroz G. Functional crosstalk of prejunctional receptors on the modulation of noradrenaline release in mesenteric vessels: A differential study of artery and vein. *Eur. J. Pharmacol.* 2011;652(1-3):33-9.

[104] Arribas SM, Daly CJ, Gonzalez MC, McGrath JC. Imaging the vascular wall using confocal microscopy. *J. Physiol.* 2007;584(Pt 1):5-9.

[105] Arribas SM, Gonzalez JM, Briones AM, Somoza B, Daly CJ, Vila E, et al. Confocal myography for the study of hypertensive vascular

remodelling. *Clinical hemorheology and microcirculation*. 2007;37(1-2):205-10.

[106] Poulsen SA, Quinn RJ. Adenosine receptors: new opportunities for future drugs. *Bioorganic & medicinal chemistry*. 1998;6(6):619-41.

[107] Fredholm BB, Cunha RA, Svenningsson P. Pharmacology of adenosine A2A receptors and therapeutic applications. *Curr. Top Med. Chem.* 2003;3(4):413-26.

[108] Cacciari B, Pastorin G, Bolcato C, Spalluto G, Bacilieri M, Moro S. A2B adenosine receptor antagonists: recent developments. *Mini reviews in medicinal chemistry*. 2005;5(12):1053-60.

[109] Birch DJ, Turmaine M, Boulos PB, Burnstock G. Sympathetic innervation of human mesenteric artery and vein. *J. Vasc. Res.* 2008;45(4):323-32.

[110] Saladin KS, editor. Anatomy and Physiology. 6th edition ed: McGraw-Hill; 2011.

[111] Okamura T, Ayajiki K, Fujioka H, Toda M, Fujimiya M, Toda N. Effects of endothelial impairment by saponin on the responses to vasodilators and nitrergic nerve stimulation in isolated canine corpus cavernosum. *Br. J. Pharmacol.* 1999;127(3):802-8.

[112] Iwatani Y, Kosugi K, Isobe-Oku S, Atagi S, Kitamura Y, Kawasaki H. Endothelium removal augments endothelium-independent vasodilatation in rat mesenteric vascular bed. *Br. J. Pharmacol.* 2008;154(1):32-40.

[113] Somoza B, Gonzalez MC, Gonzalez JM, Abderrahim F, Arribas SM, Fernandez-Alfonso MS. Modulatory role of the adventitia on noradrenaline and angiotensin II responses role of endothelium and AT2 receptors. *Cardiovasc. Res.* 2005;65(2):478-86.

[114] Thakur S, Du J, Hourani S, Ledent C, Li JM. Inactivation of adenosine A2A receptor attenuates basal and angiotensin II-induced ROS production by Nox2 in endothelial cells. *The Journal of biological chemistry*. 2010;285(51):40104-13.

[115] Ray CJ, Marshall JM. The cellular mechanisms by which adenosine evokes release of nitric oxide from rat aortic endothelium. *J. Physiol.* 2006;570(Pt 1):85-96.

[116] Ansari HR, Nadeem A, Tilley SL, Mustafa SJ. Involvement of COX-1 in A3 adenosine receptor-mediated contraction through endothelium in mice aorta. *American journal of physiology Heart and circulatory physiology*. 2007;293(6):H3448-55.

[117] Acurio J, Troncoso F, Bertoglia P, Salomon C, Aguayo C, Sobrevia L, et al. Potential role of A2B adenosine receptors on proliferation/migration

of fetal endothelium derived from preeclamptic pregnancies. *BioMed. research international*. 2014;2014:274507.

[118] Hassanian SM, Dinarvand P, Rezaie AR. Adenosine regulates the proinflammatory signaling function of thrombin in endothelial cells. *Journal of cellular physiology*. 2014;229(9):1292-300.

[119] Escudero C, Bertoglia P, Hernadez M, Celis C, Gonzalez M, Aguayo C, et al. Impaired A2A adenosine receptor/nitric oxide/VEGF signaling pathway in fetal endothelium during late- and early-onset preeclampsia. *Purinergic Signal*. 2013;9(2):215-26.

[120] Guzman-Gutierrez E, Arroyo P, Salsoso R, Fuenzalida B, Saez T, Leiva A, et al. Role of insulin and adenosine in the human placenta microvascular and macrovascular endothelial cell dysfunction in gestational diabetes mellitus. *Microcirculation*. 2014;21(1):26-37.

[121] Guzman-Gutierrez E, Westermeier F, Salomon C, Gonzalez M, Pardo F, Leiva A, et al. Insulin-increased L-arginine transport requires A(2A) adenosine receptors activation in human umbilical vein endothelium. *PLoS One*. 2012;7(7):e41705.

[122] Ishibashi Y, Duncker DJ, Zhang J, Bache RJ. ATP-sensitive K+ channels, adenosine, and nitric oxide-mediated mechanisms account for coronary vasodilation during exercise. *Circulation research*. 1998;82(3):346-59.

[123] Maddock HL, Broadley KJ, Bril A, Khandoudi N. Effects of adenosine receptor agonists on guinea-pig isolated working hearts and the role of endothelium and NO. *The Journal of pharmacy and pharmacology*. 2002;54(6):859-67.

[124] Wyatt AW, Steinert JR, Wheeler-Jones CP, Morgan AJ, Sugden D, Pearson JD, et al. Early activation of the p42/p44MAPK pathway mediates adenosine-induced nitric oxide production in human endothelial cells: a novel calcium-insensitive mechanism. *FASEB journal : official publication of the Federation of American Societies for Experimental Biology*. 2002;16(12):1584-94.

[125] Dubey RK, Gillespie DG, Jackson EK. A(2B) adenosine receptors stimulate growth of porcine and rat arterial endothelial cells. *Hypertension*. 2002;39(2 Pt 2):530-5.

[126] Bot I, de Vries H, Korporaal SJ, Foks AC, Bot M, van Veldhoven J, et al. Adenosine A(2)B receptor agonism inhibits neointimal lesion development after arterial injury in apolipoprotein E-deficient mice. *Arteriosclerosis, thrombosis, and vascular biology*. 2012;32(9):2197-205.

[127] Gonzales JN, Gorshkov B, Varn MN, Zemskova MA, Zemskov EA, Sridhar S, et al. Protective effect of adenosine receptors against lipopolysaccharide-induced acute lung injury. *American journal of physiology Lung cellular and molecular physiology.* 2014;306(6):L497-507.

[128] Li RW, Yang C, Sit AS, Lin SY, Ho EY, Leung GP. Physiological and Pharmacological Roles of Vascular Nucleoside Transporters. *J. Cardiovasc. Pharmacol.* 2011.

[129] Olsson RA, Pearson JD. Cardiovascular purinoceptors. *Physiological reviews.* 1990;70(3):761-845.

[130] Merkel LA, Lappe RW, Rivera LM, Cox BF, Perrone MH. Demonstration of vasorelaxant activity with an A1-selective adenosine agonist in porcine coronary artery: involvement of potassium channels. *J. Pharmacol. Exp. Ther.* 1992;260(2):437-43.

[131] Nakhostine N, Lamontagne D. Adenosine contributes to hypoxia-induced vasodilation through ATP-sensitive K+ channel activation. Am. *J. Physiol.* 1993;265(4 Pt 2):H1289-93.

[132] Danialou G, Vicaut E, Sambe A, Aubier M, Boczkowski J. Predominant role of A1 adenosine receptors in mediating adenosine induced vasodilatation of rat diaphragmatic arterioles: involvement of nitric oxide and the ATP-dependent K+ channels. *Br. J. Pharmacol.* 1997;121(7):1355-63.

[133] Bryan PT, Marshall JM. Adenosine receptor subtypes and vasodilatation in rat skeletal muscle during systemic hypoxia: a role for A1 receptors. *J. Physiol.* 1999;514 (Pt 1):151-62.

[134] Prentice DJ, Hourani SM. Characterisation of adenosine receptors mediating relaxation in hamster isolated aorta. *Naunyn-Schmiedeberg's archives of pharmacology*. 2000;362(4-5):427-34.

[135] Fozard JR, Baur F, Wolber C. Antagonist pharmacology of adenosine A2B receptors from rat, guinea pig and dog. *Eur. J. Pharmacol.* 2003;475(1-3):79-84.

[136] Dubey RK, Gillespie DG, Osaka K, Suzuki F, Jackson EK. Adenosine inhibits growth of rat aortic smooth muscle cells. Possible role of A2b receptor. *Hypertension.* 1996;27(3 Pt 2):786-93.

[137] Jackson EK, Gillespie DG, Dubey RK. 2'-AMP and 3'-AMP inhibit proliferation of preglomerular vascular smooth muscle cells and glomerular mesangial cells via A2B receptors. *J. Pharmacol. Exp. Ther.* 2011;337(2):444-50.

[138] Mayer P, Hinze AV, Harst A, von Kugelgen I. A(2)B receptors mediate the induction of early genes and inhibition of arterial smooth muscle cell proliferation via Epac. *Cardiovasc Res.* 2011;90(1):148-56.

[139] St Hilaire C, Yang D, Schreiber BM, Ravid K. B-Myb regulates the A(2B) adenosine receptor in vascular smooth muscle cells. *J. Cell Biochem.* 2008;103(6):1962-74.

[140] Shen J, Halenda SP, Sturek M, Wilden PA. Novel mitogenic effect of adenosine on coronary artery smooth muscle cells: role for the A1 adenosine receptor. *Circulation research.* 2005;96(9):982-90.

[141] Wiklund NP, Cederqvist B, Gustafsson LE. Adenosine enhancement of adrenergic neuroeffector transmission in guinea-pig pulmonary artery. *Br. J. Pharmacol.* 1989;96(2):425-33.

[142] Zhang GL, Miyahara H, Suzuki H. Inhibitory actions of adenosine differ between ear and mesenteric arteries in the rabbit. *Pflugers Arch.* 1989;415(1):56-62.

[143] Stoggall SM, Shaw JS. The coexistence of adenosine A1 and A2 receptors in guinea-pig aorta. *Eur. J. Pharmacol.* 1990;190(3):329-35.

[144] Goncalves J, Queiroz G. Purinoceptor modulation of noradrenaline release in rat tail artery: tonic modulation mediated by inhibitory P2Y- and facilitatory A2A-purinoceptors. *Br. J. Pharmacol.* 1996;117(1):156-60.

[145] Jackson EK, Cheng D, Tofovic SP, Mi Z. Endogenous adenosine contributes to renal sympathetic neurotransmission via postjunctional A1 receptor-mediated coincident signaling. *American journal of physiology Renal physiology.* 2012;302(4):F466-76.

[146] Vaz-da-Silva MJ, Guimaraes S, Moura D. Adenosine and the endothelium-dependent modulation of 3H-noradrenaline release in the canine pulmonary artery. *Naunyn-Schmiedeberg's archives of pharmacology.* 1995;352(6):640-5.

[147] Zhang W, Zhang Y, Wang W, Dai Y, Ning C, Luo R, et al. Elevated ecto-5'-nucleotidase-mediated increased renal adenosine signaling via A2B adenosine receptor contributes to chronic hypertension. *Circulation research.* 2013;112(11):1466-78.

[148] Carney EF. Chronic kidney disease: Renal adenosine in hypertensive CKD. *Nature reviews Nephrology.* 2013;9(6):309.

[149] Sakowicz M, Pawelczyk T. Insulin restores expression of adenosine kinase in streptozotocin-induced diabetes mellitus rats. *Molecular and cellular biochemistry.* 2002;236(1-2):163-71.

[150] Cardenas A, Toledo C, Oyarzun C, Sepulveda A, Quezada C, Guillen-Gomez E, et al. Adenosine A(2B) receptor-mediated VEGF induction promotes diabetic glomerulopathy. *Laboratory investigation; a journal of technical methods and pathology*. 2013;93(1):135-44.

[151] Figler RA, Wang G, Srinivasan S, Jung DY, Zhang Z, Pankow JS, et al. Links between insulin resistance, adenosine A2B receptors, and inflammatory markers in mice and humans. *Diabetes*. 2011;60(2):669-79.

[152] Zhou R, Chen F, Li Q, Hu DY, Liu LM. Stimulation of the adenosine A3 receptor reverses vascular hyporeactivity after hemorrhagic shock in rats. *Acta pharmacologica Sinica*. 2010;31(4):413-20.

[153] Donoso MV, Aedo F, Huidobro-Toro JP. The role of adenosine A2A and A3 receptors on the differential modulation of norepinephrine and neuropeptide Y release from peripheral sympathetic nerve terminals. *J. Neurochem.* 2006;96(6):1680-95.

[154] Banos G, Martinez F, Grimaldo JI, Franco M. Adenosine participates in regulation of smooth muscle relaxation in aortas from rats with experimental hypothyroidism. *Canadian journal of physiology and pharmacology.* 2002;80(6):507-14.

[155] Fishman P, Bar-Yehuda S. Pharmacology and therapeutic applications of A3 receptor subtype. *Curr. Top Med. Chem*. 2003;3(4):463-9.

[156] Ohta A, Sitkovsky M. Extracellular adenosine-mediated modulation of regulatory T cells. *Frontiers in immunology*. 2014;5:304.

[157] Duro E, Pallai A, Koroskenyi K, Sarang Z, Szondy Z. Adenosine A receptors negatively regulate the engulfment-dependent apoptotic cell suppression of inflammation. *Immunology letters*. 2014.

[158] Tendera M, Gaszewska-Zurek E, Parma Z, Ponikowski P, Jankowska E, Kawecka-Jaszcz K, et al. The new oral adenosine A1 receptor agonist capadenoson in male patients with stable angina. *Clinical research in cardiology : official journal of the German Cardiac Society.* 2012;101(7):585-91.

[159] Peterman C, Sanoski CA. Tecadenoson: a novel, selective A1 adenosine receptor agonist. *Cardiology in review.* 2005;13(6):315-21.

[160] Lepist EI, Damaraju VL, Zhang J, Gati WP, Yao SY, Smith KM, et al. Transport of A1 adenosine receptor agonist tecadenoson by human and mouse nucleoside transporters: evidence for blood-brain barrier transport by murine equilibrative nucleoside transporter 1 mENT1. *Drug metabolism and disposition: the biological fate of chemicals.* 2013;41(4):916-22.

[161] Muller CE, Jacobson KA. Recent developments in adenosine receptor ligands and their potential as novel drugs. *Biochim Biophys Acta.* 2011;1808(5):1290-308.

[162] Victor-Vega C, Desai A, Montesinos MC, Cronstein BN. Adenosine A2A receptor agonists promote more rapid wound healing than recombinant human platelet-derived growth factor (Becaplermin gel). *Inflammation*. 2002;26(1):19-24.

[163] Dungo R, Deeks ED. Istradefylline: first global approval. *Drugs.* 2013;73(8):875-82.

[164] Wang Z, Xuan F, Lin WH, Troyer MD, Tendolkar A, Cutler DL. Preladenant, a selective adenosine A(2)A receptor antagonist, is not associated with QT/QTc prolongation. *European journal of clinical pharmacology.* 2013;69(10):1761-7.

[165] Ellenbogen KA, O'Neill G, Prystowsky EN, Camm JA, Meng L, Lieu HD, et al. Trial to evaluate the management of paroxysmal supraventricular tachycardia during an electrophysiology study with tecadenoson. *Circulation.* 2005;111(24):3202-8.

[166] Scheiff AB, Yerande SG, El-Tayeb A, Li W, Inamdar GS, Vasu KK, et al. 2-Amino-5-benzoyl-4-phenylthiazoles: Development of potent and selective adenosine A1 receptor antagonists. *Bioorganic & medicinal chemistry.* 2010;18(6):2195-203.

[167] Hasko G, Pacher P, Deitch EA, Vizi ES. Shaping of monocyte and macrophage function by adenosine receptors. *Pharmacology & therapeutics*. 2007;113(2):264-75.

[168] Jordan JE, Zhao ZQ, Sato H, Taft S, Vinten-Johansen J. Adenosine A2 receptor activation attenuates reperfusion injury by inhibiting neutrophil accumulation, superoxide generation and coronary endothelial adherence. *J. Pharmacol. Exp. Ther*. 1997;280(1):301-9.

[169] Gomez G, Sitkovsky MV. Differential requirement for A2a and A3 adenosine receptors for the protective effect of inosine in vivo. *Blood.* 2003;102(13):4472-8.

[170] Sullivan GW, Fang G, Linden J, Scheld WM. A2A adenosine receptor activation improves survival in mouse models of endotoxemia and sepsis. *J. Infect Dis.* 2004;189(10):1897-904.

[171] Yang D, Zhang Y, Nguyen HG, Koupenova M, Chauhan AK, Makitalo M, et al. The A2B adenosine receptor protects against inflammation and excessive vascular adhesion. *The Journal of clinical investigation.* 2006;116(7):1913-23.

[172] Dhalla AK, Shryock JC, Shreeniwas R, Belardinelli L. Pharmacology and therapeutic applications of A1 adenosine receptor ligands. *Curr Top Med Chem.* 2003;3(4):369-85.

[173] Collins LE, Sager TN, Sams AG, Pennarola A, Port RG, Shahriari M, et al. The novel adenosine A2A antagonist Lu AA47070 reverses the motor and motivational effects produced by dopamine D2 receptor blockade. *Pharmacology, biochemistry, and behavior.* 2012;100(3):498-505.

[174] Baraldi PG, Tabrizi MA, Gessi S, Borea PA. Adenosine receptor antagonists: translating medicinal chemistry and pharmacology into clinical utility. *Chemical reviews.* 2008;108(1):238-63.

[175] Jung KY, Kim SK, Gao ZG, Gross AS, Melman N, Jacobson KA, et al. Structure-activity relationships of thiazole and thiadiazole derivatives as potent and selective human adenosine A3 receptor antagonists. *Bioorganic & medicinal chemistry.* 2004;12(3):613-23.

[176] Avila MY, Stone RA, Civan MM. Knockout of A3 adenosine receptors reduces mouse intraocular pressure. *Investigative ophthalmology & visual science.* 2002;43(9):3021-6.

[177] Vidal B, Nueda A, Esteve C, Domenech T, Benito S, Reinoso RF, et al. Discovery and characterization of 4'-(2-furyl)-N-pyridin-3-yl-4,5'-bipyrimidin-2'-amine (LAS38096), a potent, selective, and efficacious A2B adenosine receptor antagonist. *J. Med. Chem.* 2007;50(11):2732-6.

[178] Sun CX, Zhong H, Mohsenin A, Morschl E, Chunn JL, Molina JG, et al. Role of A2B adenosine receptor signaling in adenosine-dependent pulmonary inflammation and injury. *The Journal of clinical investigation.* 2006;116(8):2173-82.

[179] Mustafa SJ, Nadeem A, Fan M, Zhong H, Belardinelli L, Zeng D. Effect of a specific and selective A(2B) adenosine receptor antagonist on adenosine agonist AMP and allergen-induced airway responsiveness and cellular influx in a mouse model of asthma. *J. Pharmacol. Exp. Ther.* 2007;320(3):1246-51.

[180] Bar-Yehuda S, Rath-Wolfson L, Del Valle L, Ochaion A, Cohen S, Patoka R, et al. Induction of an antiinflammatory effect and prevention of cartilage damage in rat knee osteoarthritis by CF101 treatment. *Arthritis and rheumatism.* 2009;60(10):3061-71.

[181] Bar-Yehuda S, Luger D, Ochaion A, Cohen S, Patokaa R, Zozulya G, et al. Inhibition of experimental auto-immune uveitis by the A3 adenosine *receptor agonist CF101. International* journal of molecular medicine. 2011;28(5):727-31.

[182] Cohen S, Stemmer SM, Zozulya G, Ochaion A, Patoka R, Barer F, et al. CF102 an A3 adenosine receptor agonist mediates anti-tumor and anti-inflammatory effects in the liver. *Journal of cellular physiology.* 2011;226(9):2438-47.

[183] Doherty GJ, McMahon HT. Mechanisms of endocytosis. *Annual review of biochemistry.* 2009;78:857-902.

[184] Sorkin A, von Zastrow M. Endocytosis and signalling: intertwining molecular networks. *Nature reviews Molecular cell biology.* 2009;10(9):609-22.

[185] Katzmann DJ, Babst M, Emr SD. Ubiquitin-dependent sorting into the multivesicular body pathway requires the function of a conserved endosomal protein sorting complex, ESCRT-I. *Cell.* 2001;106(2):145-55.

[186] Haglund K, Dikic I. The role of ubiquitylation in receptor endocytosis and endosomal sorting. *Journal of cell science.* 2012;125(Pt 2):265-75.

[187] Raiborg C, Stenmark H. The ESCRT machinery in endosomal sorting of ubiquitylated membrane proteins. *Nature.* 2009;458(7237):445-52.

[188] Grube K, Rudebusch J, Xu Z, Bockenholt T, Methner C, Muller T, et al. Evidence for an intracellular localization of the adenosine A2B receptor in rat cardiomyocytes. *Basic research in cardiology.* 2011;106(3):385-96.

[189] Alba M, Silvia G, Lanfranco C, Ilaria B, Grazia RM, Carla E, et al. Adenosine A(1) receptors contribute to mitochondria vulnerability to pro-oxidant stressors. *Mitochondrion.* 2010;10(4):369-79.

In: Adenosine Receptors
Editor: Kasandra Warrick

ISBN: 978-1-63463-454-0

Chapter III

Insights on the Role of Adenosine and Adenosine Receptors on Melanoma: New Therapeutic Strategies

A. S. Soares[1,2], J. B. Sousa[1,2], C. Diniz[1,2] and P. Fresco[1,2]
[1]REQUIMTE/FARMA, Department of Drug Sciences, Laboratory of Pharmacology, University of Porto – Faculty of Pharmacy, Porto, Portugal
[2]MedInUP – Center for Drug Discovery and Innovative Medicines, University of Porto, Porto, Portugal

Abstract

The role of adenosine and adenosine receptors in tumourigenesis has been largely investigated. Studies suggest a contradictory role of adenosine in viability of normal and cancer cells, ascribed to the stimulation of adenosine receptors subtypes, named A_1, A_{2A}, A_{2B}, and A_3, which couple to different signal transduction pathways and can be co-expressed in the same cell.

Several studies have suggested a crucial role for A_3 adenosine receptors on cancer cells' cytotoxicity, improving, by opposition, normal cells proliferation. This dual effect, coupled with the fact that these receptors are overexpressed in several tumours and expressed at low levels in normal tissues, suggested A_3 agonists as promising cancer therapeutic strategies. However, time has proved this to be an oversimplified view. In fact, activation of adenosine A_3 receptors can lead both to malignant cell proliferation and death, depending on agonist's concentration. Moreover, A_3 agonists can mediate cytotoxicity independently of adenosine A_3 receptor activation.

Melanoma is considered one of the most aggressive malignant tumours, due to its high metastatic potential and resistance to current treatment modalities. Patients with metastatic melanoma present poor prognosis, with an overall survival of 8-18 months. Melanoma incidence is increasing faster than any other solid tumour accounting for a global annual incidence of about 160 thousands new cases.

This chapter intends to integrate our research work and the current knowledge concerning the role of adenosine/adenosine receptors in the complex interplay of proliferation, cell death mechanisms and metastatic progression of melanoma.

Our group has shown adenosine receptor subtypes expression, on different metastatic melanoma cell lines. Their activation did not affect cell proliferation of human A375 and mouse K1735-M2 cells. However, proliferation of human C32 cells was increased, after activation of A_3 adenosine receptor, by micromolar concentrations of adenosine or nanomolar concentrations of the selective A_3 agonist, Cl-IB-MECA. By opposition, Cl-IB-MECA, at micromolar concentrations, caused marked cytotoxicity, through A_3-independent mechanisms. A major role for adenosine A_3 receptors in C32 cell proliferation induced by an adenosine metabolite, inosine, through PLC-PKC-MEK1/2-ERK1/2 pathway activation was demonstrated. Inosine also enhanced proliferation of A375 cells and stimulated metastatic processes evaluated in both cell lines (migration, adhesion, invasion and colony-formation), through adenosine A_3 receptor activation.

The combination Cl-IB-MECA and paclitaxel was investigated and found to induce multiple mechanisms of cell death and inhibit metastatic processes. This combination emerges, therefore, as a promising new therapeutic strategy for human metastatic melanoma, potentially overcoming chemotherapy multiresistance and improving patient prognosis.

Introduction

Evidence suggests that complex interactions of tumour cells with the tumour microenvironment may be related with secretory products released by these cells, which appear to be fundamental for tumourigenesis [1]. Adenosine

(ADO), the end product of adenosine triphosphate (ATP) hydrolysis, is being pointed out as an autocrine and paracrine regulatory factor which can rapidly increase its concentration in response to pathophysiological conditions, and accumulates in the tumour microenvironment [2, 3]. Indeed, alterations in purine metabolism have been found to facilitate ADO production and contribute to tumour development and progression [4]. Moreover, increased levels of ADO also contribute to angiogenesis that is essential for tumour growth [5], including melanoma [6]. Furthermore, it is currently accepted that ADO may regulate tumour growth through both receptor-mediated and receptor-independent pathways [7]. Extracellular ADO acts mainly through the activation of four subtypes of adenosine receptors (AR), A_1, A_{2A}, A_{2B}, and A_3, members of the G protein-coupled receptor superfamily, and it is considered a full agonist for all AR [3]. Alternatively, ADO may be rapidly transported into the cell by nucleoside transporters [8]. Once inside the cell, adenosine can be converted to adenosine monophosphate (AMP) via adenosine kinase (AK) or to inosine (INO) via adenosine deaminase (ADA) [9], which can also bind to AR and initiate intracellular signalling events or affect cell function *via* receptor-independent pathways [10].

Metastatic melanoma is considered one of the most aggressive malignant tumours, representing the deadliest form of skin cancer [11]. This serious disease, derived from transformed epidermal melanocytes, presents an alarming increase in incidence rates [12]. Although melanoma is less common than other types of skin cancers, the incidence of cutaneous melanoma is rising faster than any other solid tumour [13], accounting for a global incidence of about 160 thousands new cases each year [14]. This increased incidence may be a consequence of more people chronically exposed to natural/artificial UV radiation and/or as a result of higher detection rates associated with enhanced public awareness [15]. The ability of melanoma to metastasize results in early and general spread and very high mortality rates [11]. Resistance to chemotherapy and radiotherapy [16] is the main reason for patients with metastatic melanoma to have an overall survival of less than two years [11]. The current absence of an effective pharmacological approach could be, at least in part, due to an incomplete understanding of the pathophysiological mechanisms involved in melanoma cells proliferation and, ultimately, tumour initiation and progression. Abnormal activation of pathways that regulate the proliferation of melanocytes is associated to melanoma progression [17]. Mitogen-activated protein kinase (MAPK) and phosphoinositide 3-kinase (PI3K) pathways play a key role in cell proliferation, being constitutively activated in most melanomas [14, 18].

While primary melanoma is highly curable with surgery, patients with metastatic disease carry the worst prognosis [15]. One of the major problems of metastatic melanoma has been the lack of efficient treatments [16] which could be related to an incomplete understanding of mechanisms that regulate tumour initiation and metastatic progression. Indeed, it is the most treatment-resistant human cancer [19], remaining one of the greatest challenges in oncotherapy. It is accepted that the limited success of most current therapeutic strategies used in metastatic melanoma is due, at least in part, to cancer cells ability in escaping apoptosis, thus leading to drug resistance [20]. Drugs that activate pathways overcoming melanoma's apoptosis resistance and induce cell death have, therefore, great potential for melanoma therapeutic interventions.

Our research has been focused both in understanding the role of ADO, its metabolism and its receptors (particularly A_3AR) on the mechanisms of cell proliferation, death and metastatic progression in human melanoma cells. This chapter describes these advances, integrating them with the current knowledge. A particularly important contribution of our research has been the recent proposal of a new promising cytotoxic combination for melanoma treatment, Cl-IB-MECA with low doses of the chemotherapeutic agent paclitaxel, which has been proved to act through the induction of multiple mechanisms of cell death and blockade of several metastatic pathways involved in melanoma progression. This combination seems to be able to prevent melanoma multiresistance to chemotherapy and stop its progression, therefore, potentially improving melanoma treatment.

1. Adenosine

ADO is a purine nucleoside that plays a central role as a structural element of nucleic acid and in the energy metabolism of all living organisms. Extracellular ADO acts as a signalling molecule, mediating numerous physiological and metabolic effects [21] being considered a major local regulator of tissue function, especially when energy supply cannot fulfil cellular energy demand [22].

ADO accumulation during ischemia and inflammation protects tissue from injury [23] by several mechanisms that include stimulation of glycogen breakdown to provide glucose for ATP production via anaerobic glycolysis,

[21], increasing the ratio of oxygen supply to demand by causing vasodilation, reducing inflammation and promoting angiogenesis [24, 25].

1.1. Formation and Metabolism

ADO is generated and metabolized through stepwise dephosphorylation of ATP. ATP and ADP are converted by ecto-ATP apyrase (CD39) to AMP, through a process activated by hypoxia [21]. Dephosphorylation of AMP is mediated, intracellularly, by cytosolic-5′-nucleotidase and extracellularly by membrane-bound ecto-5′-nucleotidase (CD73), respectively. Adenosine can also be produced (marginally) from intracellular hydrolysis of S-adenosylhomocysteine [24]. ADA is widely distributed in many cells and tissues where it metabolizes adenosine into inosine and hypoxanthine. AK catalyses the formation of AMP from adenosine in order to reload ATP cellular stores [21, 24]. Due to its hydrophilic nature, cellular uptake and release of adenosine is accomplished by two main groups of membrane nucleoside transporters (NT). Concentrative nucleoside transporter family (SLC28) is found in specialized epithelial tissue and it mediates active Na^+-dependent nucleoside transport. Passive nucleoside transport processes are ubiquitous and are mediated by members of the equilibrative nucleoside transporter (ENT) family (SLC29) [26]. Four ENT isoforms have so far been identified in human and rodent genomes. Inhibition of ENTs by dipyridamole prevents the reuptake of adenosine into endothelial cells, increasing its extracellular concentration and enhancing AR activation [21, 24].

Extracellular ADO concentrations can increase by increasing the activities of nucleotidases (major pathway) and AKs (minor pathway) along with their substrates (such as AMP). In addition, inhibition of ENTs can also lead to increased extracellular ADO [24]. Exposure to a hypoxic environment (namely in tumour microenvironment) can have these effects on nucleotidases and AKs, leading to a net increase in the production of ADO [23]. Hypoxia has also been reported to downregulate ENT1 expression in different mouse tissues [27], thereby increasing extracellular concentrations of ADO.

ADO accumulates in tissues when oxygen demand exceeds oxygen supply, as in tumour microenvironment [1]. Physiological extracellular ADO concentrations have been estimated to be approximately 30-300 nM [28]. However, as these are dependent of energy consumption it is not surprising that under metabolic stress conditions, such as hypoxia and ischemia where breakdown of ATP is enhanced, ADO extracellular levels can reach concentrations as higher as 30 μM [29]. The concentration of the ADO metabolite, INO, follows the fluctuations of ADO concentration, resulting in similar levels of ADO and INO most of the time [30]. *In vivo* studies revealed that ADO concentrations in the extracellular fluids of solid tumours, in mice, ranged from 0.2 to 2.4 μM, concentrations 10- to 20-fold higher than those found in adjacent subcutaneous tissues. Moreover, treatments with ADA and AK inhibitors were able to increase ADO levels in tumours up to 13 μM [31]. Studies have shown that adenosine levels are highest in the tumour microenvironment which is severely hypoxic. ADO produced under such conditions can definitely activate ARs present on the tumour cells [32, 33].

1.2. Adenosine Receptors

Most effects of extracellular ADO are mediated by activation of four known ARs subtypes, namely the A_1, A_{2A}, A_{2B} and A_3AR, which are surface G-protein-coupled receptors with distinct tissue-specific distributions [25, 28] These receptors, differ in their affinity for ADO, in the G proteins recruited, and finally in the activation of downstream signalling pathways of target cells. A_1, A_{2A} and A_3ARs display high affinity while A_{2B}AR displays low affinity for ADO [34]. A_1AR and A_3AR, couple to G_i proteins and inhibit the enzyme adenylyl cyclase (AC), thereby decreasing the levels of intracellular cyclic AMP (cAMP) [28]. However, they also activate other intracellular signal transducers, such as PKC, PI3K, and ERK1/2 [35]. Additionally, A_3AR is also able to signal through the PLC pathway via G_q proteins [36]. A_{2A}AR and A_{2B}AR, couple to G_s proteins and activate AC, leading to the increase of cAMP levels. This increase results in the activation of the protein kinase A (PKA) pathway [28]. Moreover, A_{2A}AR has been described to activate G protein-independent pathways [37] and A_{2B}AR is also able to couple G_q, as well as to activate other pathways, such as MAPK and PI3K [35]. It seems conceivable that MAPK and PI3K signalling and, therefore, cell proliferation might be amenable to manipulation through specific AR in tumour cells.

1.3. Adenosine Receptors in Cancer

Effects of ADO are pleiotropic, being the cellular response to this nucleoside dependent on the expression of the different AR subtypes on each tissue. They can be co-expressed in the same cell, activating different signal transduction pathways.

ARs have been actively studied as potential therapeutic targets in several disorders, including cancer (reviewed in [38]) and have been found to be upregulated and differentially expressed on tumour cell lines and tumour samples [33].

Increasing evidence indicates that ADO is an autocrine and paracrine regulatory factor that can rapidly increase its concentration in response to pathophysiological conditions, and therefore accumulates in the tumour microenvironment [2, 3]. Indeed, alterations in purine metabolism were found to facilitate ADO production, thereby contributing to tumour development and progression [4]. Moreover, increased levels of ADO also contribute to angiogenesis that is essential for tumour growth [5]. ADO may regulate tumour growth through both receptor-mediated and receptor-independent pathways [7]. Extracellular ADO acts mainly through the activation of AR, and it is considered a full agonist for all AR [3]. Moreover, reduction of extracellular levels of ADO, by an increase in ADA activity, has been also suggested to reduce tumour promoting functions [39]. Nevertheless, ADO showed growth-inhibitory effects, and anti-apoptotic or even pro-apoptotic effects, depending on the tumour type (reviewed in [3]), suggesting that activation of AR subtypes involved in tumour progression remains largely unclear. Although both *in vitro* studies and animal models have confirmed that targeting the AR has tremendous potential for treating cancer [2], the translation of these results into clinical practice will require a more thorough understanding of how ADO regulates cancer.

High levels of A_1AR are expressed in breast cancer tissues, compared to normal breast tissues [40]. A_1AR activation was shown to reduce cell proliferation by inhibiting the G_1 checkpoint in the cell cycle [40]. Increased A_1AR expression was also observed in the peritumoral regions (activated astrocytes and microglia) of experimental F98 glioma cells and showed to suppress glioblastoma [41]. In addition, activation of A_1AR on colon cancer cells increased apoptosis, by activating caspases [42].

Studies have also demonstrated an A_{2A}AR anti-tumour activity in A375 melanoma [43] and Caco-2 colon [44] cancer cells, the latter mediated by activation of caspases. In contrast, A_{2A}AR activation promoted growth of breast cancer MCF-7 breast cancer cells [45] and increased angiogenesis by inducing VEGF expression [46], suggesting that A_{2A}AR antagonists could be useful in cancer treatment.

$A_{2B}AR$ activation is also linked to angiogenesis and promotes growth and tumour metastasis [47]. Moreover, $A_{2B}AR$ knockout mice demonstrated reduced tumour growth and longer survival times after being administered Lewis lung carcinoma, comparatively to wild-type mice [48].

U87MG human glioblastoma cells revealed transcripts for A_1, A_{2A} and $A_{2B}ARs$, but no transcript for A_3AR [49]. However, other authors were able to detect A_3AR in these cells [50]. A_3AR was also found to be highly expressed in colon and breast carcinoma tissues as compared to non-neoplastic or normal tissues [51]. A role of A_3AR in mediating anti-tumour actions *in vitro* and *in vivo* has also been demonstrated. In animal models, A_3AR agonists were reported to inhibit growth of melanoma [52] and colon [53, 54] cancers.

A_3AR overexpression occurs in several types of cancers [51], when compared to normal adjacent tissues. Primary and metastatic tumour tissues were found to highly express A_3AR, indicating that high receptor expression is a characteristic of solid tumours, suggesting A_3AR as a potential target for tumour growth inhibition [51]. Interestingly, expression of A_3AR was found to be low in most body tissues, while it is highly expressed in tumour cell lines [55-57].

Treatment of mice inoculated with human HCT-116 or murine CT-26 colon carcinoma cells with A_3AR agonists reduced liver metastasis formation [53]. Other authors ascribed the A_3AR agonists anti-tumour activity to increase in natural killer (NK) cell activity, which promotes killing of tumour cells [58]. Although previous studies demonstrated that A_3AR agonists, such as IB-MECA or Cl-IB-MECA, at micromolar concentrations, induced apoptosis in different tumour cells, this was not true for all cells tested. Cl-IB-MECA (concentrations $\geq$ 10 μM) led to apoptosis in leukemic (30 μM) and thyroid carcinoma cells through an A_3AR-independent mechanism [59-61], suggesting the possibility that these effects are most likely dependent on the intracellular level of Cl-IB-MECA. Moreover, Cl-IB-MECA is currently being evaluated in clinical trials (phase I and phase II) on patients with hepatocellular carcinoma, showing no serious drug-related adverse events or dose-limiting toxicities and good oral bioavailability [62].

1.3.1. Adenosine Receptors in Melanoma

Studies have demonstrated that AR activation in melanoma can lead to both pro- and anti-tumorigenic effects. At present, research results indicate that, in melanoma, activation of A_3AR has been shown to have a very important role in controlling melanoma growth, although there are also some studies concerning other ARs (reviewed in [63]).

1.3.1.1. A_3AR and Melanoma

Targeting A_3AR by ADO or synthetic agonists has been described to cause differential effects on tumour or normal cells, inhibiting cell growth of various tumour types, including melanoma, while favouring growth of normal cells [64].

The pharmacological and biochemical profiles typical of A_1, A_{2A}, A_{2B} and A_3 AR subtypes was presented on human melanoma cells, for the first time, in A375 cells, showing that A_1 and A_3 mRNAs were less expressed than A_{2A} and A_{2B} mRNAs [55]. The ability of IB-MECA, a selective A_3AR agonist, to decrease the levels of PKA, a downstream effector of cAMP, and AKT, leading to cell growth suppression, was described in mouse B16-F10 melanoma cells [64]. Moreover, in melanoma tumour lesions derived from IB-MECA-treated mice, the expression level A_3AR and its activation inhibited PKA and AKT signalling proteins, resulting in tumour growth suppression [52]. Cordycepin (3'-deoxyadenosine) was also able to inhibit the growth of another mouse B16-BL6 melanoma cells type, through the stimulation of A_3AR [65]. Oral administration of Cl-IB-MECA, also considered a selective A_3AR agonist, to melanoma-bearing mice suppressed the development of melanoma lung metastases, causing synergistic anti-tumour effect when in combination with cyclophosphamide while preventing the cyclophosphamide-induced myelotoxic effects [66].

Activation of A_3AR with Cl-IB-MECA, on human A375 melanoma cells, resulted in PI3K-dependent phosphorylation of AKT, leading to the reduction of basal levels of ERK1/2 phosphorylation, which in turn inhibited cell proliferation [67]. However, ADO, the natural ligand, displays contradictory effects on A375 cells: it improves cell proliferation through A_3AR activation, while it may also induce apoptosis and impair cell survival via $A_{2A}AR$ signalling pathway, which involves PKC and MAPK (43). The blocking of A_3AR, using the selective A_3AR antagonist MRE3008F20, decrease VEGF secretion in melanoma cells treated with etoposide and doxorubicin, presenting the possibility of improving the ability of chemotherapeutic drugs to block angiogenesis [68].

2. Melanoma

Melanoma is the most serious form of skin cancer and results from the proliferation and malignant transformation of melanocytes located in basal layer of the epidermis, which may evolve from preexisting melanocytic nevi or as a new process. Melanocytes are originary from the neural crest, from which they migrate, during the embryonic period, to the skin, uvea, leptomeninges and mucous membranes – locations where melanoma can occur. After migration, melanocytes remain undifferentiated until stimulated by ultraviolet radiation (UVR), differentiating into cells producing melanine [69, 70]. In the transformation process both environmental and host related factors are involved [11, 71]. Endogenous risk factors like previous family history of melanoma, and multiple dysplastic or benign nevi, blond or red hair, and pale skin have revealed a strong association with melanoma development [11]. Nevertheless, exogenous risk factors including long term sun exposure, photochemotherapy or chemical exposure are also reported to favour the development of melanoma [72]. In individuals with a tendency to develop large numbers of nevi, melanomas are initiated after modest solar radiation exposure, after which host factors drive further progression; on the contrary, in individuals with little or no tendency to develop nevi, repeated exposure to the sun seems crucial for melanoma development [10]. This could justify the different mutation patterns found in skin areas chronically or non-chronically exposed to solar radiation [73]. Nevertheless, approximately 75% of melanomas arise from normal skin in the absence of pre-existing nevi [74].

Solar UV radiation has been linked to skin cancer, both in non-melanoma (transformed keratinocytes) and melanoma (transformed melanocytes) skin cancers [75]. Skin malignant changes promoted by UV solar radiation (i) cause DNA mutagenic effects; (ii) increase in the production of growth factors by skin cellular constituents; (iii) reduce the cutaneous immune defences; and (iv) promote reactive oxygen species (ROS) formation [76]. Although epidemiological studies have linked solar UV radiation to melanoma [77], the specific contribution of UVB and UVA radiation exposure in the risk of melanoma is controversial. Nevertheless, evidence suggests that both UVA and UVB radiation act together to promote the development and progression of melanoma [78, 79]. UVB, the minor but the most damage solar light component, causes directly DNA damage on cells of skin epidermal basal layer, since UVB wavelengths are within the DNA absorption spectrum [80]. UVA wavelengths (more abundant) penetrate into the epidermis and dermis of the skin, generating ROS that can damage cellular components, including DNA [81]. Skin has endogenous defence systems that limit the potential damage caused by UV radiation, such as melanin, antioxidant defences, and DNA repair enzymes [80]. However, these endogenous photoprotective mechanisms are not always enough to neutralize UV radiation adverse effects.

As such, melanocytes that have accumulated genetic abnormalities may develop melanoma. These can promote malignant phenotype, including cell proliferation and impaired apoptosis, stimulation of angiogenesis, escape from the immune system, tumour invasion, and finally metastasis [76]. Malignancy hallmarks including rapid and uncontrolled proliferation, invasion, and colonization to distant organs can be attributed, in part, to deregulation of interactions between neoplastic cells and cellular microenvironment [82].

Following malignant transformation, melanoma cells become autonomous and no longer respond to keratinocytes [83]. Development of melanoma metastasis starts with cell migration from the primary tumour, followed by re-adhesion and tissue invasion, with complex interactions between tumour cells and their environment [84]. Proteolytic degradation of basement membrane and ECM represents a critical step for melanoma invasion and metastasis, since it displaces the surrounding matrix required for cells to move into tissues [85]. In addition, anchorage-independent growth (colony formation) is also important for malignant cell transformation as this ability identifies tumours with metastatic potential [86]. Newly colonies of melanoma cells dramatically increase the consumption of oxygen and nutrients, eventually leading to cell starvation and hypoxia at the tumour site [87]. These increasing demands induce the development of additional vasculature in order to increase blood supply, triggering angiogenesis, namely by increasing ADO levels [5]. Indeed, the striking correlation between the degree of angiogenesis and metastatic melanoma colonies strongly indicates a dependency of newly formed colonies of melanoma cells on angiogenesis [6].

2.1. Signalling Pathways Involved in Melanoma Progression

Disruption in crucial signalling pathways that regulate normal melanocyte function is strongly linked to melanoma development and progression [17, 88]. Several signalling pathways were found to be constitutively activated in melanoma. The mitogen-activated protein kinase (MAPK) and phosphoinositide 3-kinase (PI3K)-AKT signalling pathways are activated in melanoma and appear to play a major role in cell proliferation, and melanoma development and progression [89], since these pathways were found to be constitutively activated in most melanomas [14, 18].

2.1.1. Mitogen-Activated Protein Kinase Pathway

Aberrant MAPK [Ras–Raf–mitogen-activated protein kinase kinase (MEK)–extracellular signal-regulated kinase (ERK)] pathway activation is being implicated in the regulation of cancer cell behaviour, namely cellular proliferation, differentiation, and survival [90]. Briefly, stimulation of Ras causes activation of Raf, which then phosphorilates MEK, leading to activation of ERK.

Activation of the MAPK pathway is a frequent and early event in melanoma [91]. During the process of malignant transformation, melanocytes acquire the ability to secrete autocrine and paracrine growth factors [92]. Many of these factors contribute to the increased ERK activity in melanoma through direct activation of the Ras-Raf pathway [18]. Nevertheless, this may also occurs *via* phospholipase C (PLC)-protein kinase C (PKC), which can induce direct activation of Raf, leading to MEK1/2 activation (reviewed in [93]). Moreover, ERK1/2 activation involved in cellular proliferation was also reported to be dependent of G protein-PKC-MEK1/2 signalling [94].

Mutations of N-Ras or B-Raf, autocrine and paracrine growth factors, and adhesion receptor activation represent some of the multiple mechanisms through which Ras-Raf-MEK-ERK signalling pathway is activated in melanoma [18]. Extracellular signals generated through the interaction of ligand-receptor are transmitted to the interior of the melanoma cells, activating downstream signalling cascades. Finally, ERK1/2 either phosphorylates cytoplasmic targets or migrates to the nucleus, where it phosphorylates and activates transcription factors [95].

Evidence indicates that MEK1/2 inhibitors can block ERK activity of human melanoma cell lines, leading to melanoma cell cycle arrest [96]. Therefore, MAPK pathways seem essential for cellular proliferation.

2.1.2. Phosphatidylinositol-3-Kinase Pathway

Constitutive activation of the PI3K-AKT pathway can trigger a cascade of responses, from cell growth and proliferation to survival and motility, which drive to tumour progression [97]. Briefly, PI3K is activated by receptor protein tyrosine kinases or receptors coupled with G-proteins. The substrate for class I PI3K is phosphatidylinositol-4,5-biphosphate, which leads to the production of phosphatidylinositol-3,4,5-triphosphate (PIP3). Then, PIP3 recruits signalling proteins, namely AKT (also named protein kinase B) inducing conformational changes. Active AKT then translocates to the nucleus and mediates the activation and inhibition of several targets, resulting in cellular survival and proliferation [18].

Although in melanoma PI3K itself is rarely mutated [98] or overexpressed [99], activation of downstream signalling components, namely AKT, have been implicated in melanoma progression [100]. As such, interaction of extracellular signals with the respective receptor is transmitted to the interior of the melanoma cells, activating downstream cascades, with AKT activation being crucial for melanoma progression, through regulation of adhesion molecules [18]. Constitutive activation of AKT was observed in melanoma cell lines and melanoma samples in different progression stages [101]. Moreover, increased phospho-AKT expression is also associated with melanoma progression and a worse patient survival [102].

2.2. Therapies for Metastatic Melanoma

Metastatic melanoma is highly resistant to chemotherapy, radiotherapy, hormonal therapy and immunotherapeutic approaches (reviewed in [103]). In the last few decades overall survival has slightly improved, mainly due to earlier diagnosis. Nevertheless, no significant impact on survival has been demonstrated, despite improvements in response rates, with new drugs or new combinations of drugs [16].

Dacarbazine is one of the most effective chemotherapeutic agents for metastatic melanoma and remains the classical mono-chemotherapeutic regimen used for melanoma. This alkylating DNA agent, used as a single agent shows response rates of approximately 20% (104). Although it has low activity as monotherapy, intravenous dacarbazine has remained the basis for many chemotherapeutic combinations, including cisplatin and tubular toxins. However, these combinations had limited impact on overall survival and present increased toxicity [105].

Anti-microtubular agents, such as paclitaxel (PXT; natural product-based taxane) that act through stabilisation of polymerised microtubules, have been reported to have modest activity in patients with metastatic melanoma (16.4%) but comparable with that of dacarbazine, cisplatin, or interleukin-2 (IL-2) [106]. However, in another combination an overall response rate of 24% was reported in phase II trial with PXT and tamoxifen [107]. Interestingly, a phase I/II trial of sorafenib (non-selective B-Raf inhibitor) administered in combination with carboplatin and PXT demonstrated preliminary antitumor activity in metastatic melanoma with a favourable safety profile [108]. Moreover, PXT is being used on a phase II trial in combination with carboplatin and axitinib, a selective second-generation inhibitor of vascular endothelial growth factor (VEGF) receptors, on metastatic melanoma patients, although final data is not yet published (Clinicaltrials.gov identifier: NCT01174238).

High dose IL-2, an immunodulatory agent, seems to benefit some patients with metastatic melanoma, presenting an overall response of 16%, with complete and partial responses rates of 6% and 10%, respectively [109]. IL-2 mediates its effects by binding to IL-2 receptors, which are expressed by lymphocytes, the cells that are responsible for immunity. Unfortunately, these more effective response rates were also associated with greater toxicities [110].

Ipilimumab, a monoclonal antibody that blocks cytotoxic T lymphocyte-associated antigen-4, being a negative regulator of T-cells [111], has recently been approved by the FDA as a single agent therapy for the treatment of late stage melanoma. With the approval of ipilimumab, the role of high dose IL-2 remains in question. While the clinical benefit of high dose IL-2 is roughly similar to ipilimumab, it presents higher toxicity and lower tolerability [112]. Nevertheless, ipilimumab has relatively modest overall survival improvement, and toxicities associated to this therapy can be severe, long-lasting, or both, although most being reversible with appropriate treatment [111].

Novel therapies which target specific alterations associated with melanoma are also appearing [113]. The first selective B-Raf inhibitors, vemurafenib and dabrafenib, demonstrated an impressive single agent activity against melanoma, in clinical trials [114, 115]. However, both also showed significant skin-related toxic effects (squamous cell carcinomas) [114, 115], with patients becoming resistant to treatment, which has been attributed to reactivation of the MAPK pathway and enhanced tumour growth through c-Raf [116]. Therefore, combination therapy with MEK inhibitors may be preferable, as all isoforms of both Ras and Raf signal through MEK, and this would lead to complete inhibition of the MAPK pathway. Phase I and II trials of combined treatment with dabrafenib and trametinib (a selective MEK inhibitor) have shown increased progression free survival, without significant reduction in the appearance of squamous cell carcinoma [117]. Tipifarnib (R115777), a Ras inhibitor, has been shown to have anti-tumour effects in both human cell cultures and in vivo model of melanoma [118]. However, this drug never reached clinical trials as a single-agent, and when used in combination with sorafenib (a non-selective B-Raf inhibitor), it did not significantly improved sorafenib results [119].

Other targets being investigated as possible therapeutic targets for melanoma include inhibitors of the PI3K pathway. However, the dual

PI3K/mammalian target of rapamycin (mTOR) inhibitor, PI-103, promoted immunosuppression, *in vivo* tumour growth and increases in survival of sorafenib-treated melanoma cells [120], whereas sorafenib in combination with temsirolimus (mTOR inhibitor), tested in a phase II trial, did not show sufficient activity to justify further use [119].

The limited efficacy of current therapies and their significant side effects reinforce the need for the development of novel treatment strategies and redirect investigations towards new therapeutic targets.

3. New Insights

In this section, we detail the most important results of our research in the context of gathering new insights on the role of the ADO-AR system in the complex interplay of proliferation, death and metastatic progression of melanoma cells as clues for the proposal of new therapeutic approaches for this deadliest form of skin cancer.

3.1. Characterization of Adenosine Receptor Expression in Melanoma Cell Lines

Our group studied, for the first time, the expression of all subtypes of AR (A_1, A_{2A}, A_{2B} and A_3) on human C32 and mouse K1735-M2 melanoma cells, by immunocytochemistry, and confirmed their presence on human A375 melanoma cell line [121]. The A_3AR seems to be poorly expressed in all cell lines [121], which is in agreement with previous work showing that A_3 mRNAs are less abundant than A_{2A} and A_{2B} mRNAs in human A375 melanoma cells [55]. Moreover, although melanoma and other tumours express high levels of A_3AR [51], this fact is not true for all types of cancer cells, as human MCF-7 breast cancer cells have undetectable AR expression, whereas human MDA-MB-231 breast cancer cells only express high levels of A_{2B}AR [122].

We also performed a pharmacological characterization showing that the activation of A_1AR, A_{2A}AR, and A_3AR did not interfere on cell proliferation of human A375 and mouse K1735-M2 melanoma cells [121]. However, on human C32 melanoma cells, A_3AR activation elicited by micromolar concentrations of ADO, or by nanomolar concentrations of the selective A_3AR agonist Cl-IB-MECA, induced cell proliferation. These findings suggest that activation of AR subtypes on melanoma cells is dependent on species and gender. In fact, C32 and A375 cells were collected from male and female, respectively, and different melanoma metastatic patterns were found on male and female patients [123]. The aforementioned results highlight the fact that caution should be taken in data extrapolation from mice to humans, and generalization of AR activation in human disease should be avoided. Nevertheless, these findings suggest that high expression of A_3AR is not essential to obtain effects. In fact, it seems that although being poorly expressed, it is the only receptor subtype that mediates ADO effects on human C32 melanoma cells.

3.2. Inosine Strongly Enhances Human C32 Melanoma Cells Proliferation through PLC-PKC-MEK1/2-ERK1/2 and PI3K Pathways

Evidence suggests that an increase in adenosine deaminase (ADA) activity can cause a reduction in the extracellular levels of ADO and, therefore, reduce tumour promotion [39]. The hypothesis that degradation of ADO would prevent melanoma cells proliferation was evaluated by our group, since ADO induced proliferative effects on C32 melanoma cells [121]. The effects of ADO and its degradation (addition of ADA) were further studied, with the associated mechanisms being clarified. One major and surprising finding was that inosine (INO) caused C32 cell proliferation more pronouncedly than ADO itself, those effects being mediated by A_3AR activation [124]. Moreover, reduction of cell proliferation by AOPCP (an ecto-5'NTase inhibitor) was demonstrated, suggesting that increased levels of AMP, in contrast to ADO, have an inhibitory effect on C32 cell proliferation [124]. Similar results obtained with AOPCP were reported in glioma [125] and breast [126] cancer cells. These findings highlight the complexity of the effects of ADO metabolism, suggesting that ADO degradation, through addition of ADA, is not a successful approach in preventing cell proliferation in this model. Indeed, this strategy may even worsen the clinical condition, since not only ADO, but also INO increases melanoma cell proliferation. Studies reporting increased levels of ADA in serum of patients with bladder [127], breast [128] and lung [129] cancer could be linked to our finding, suggesting a relevant role of the metabolic product of ADO in cancer. Attempts were also made in order to clarify the intracellular mechanisms by which INO mediates cell proliferation. Constitutive activation of PLC-PKC-MEK1/2-ERK1/2 and PI3K pathways was found in C32 melanoma cells and INO showed to increase cell proliferation through further increments of these pathways [124] (Figure 1).

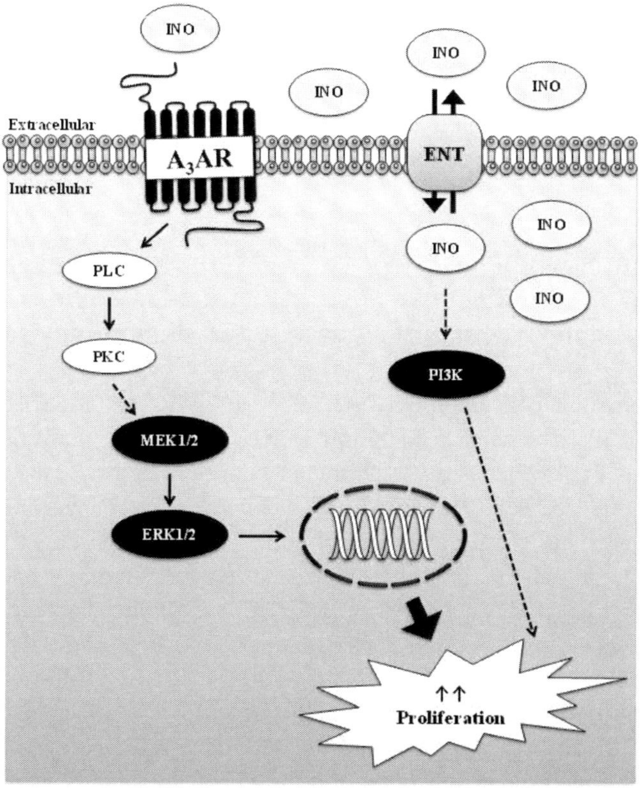

Figure 1. Proposed signalling pathways involved in proliferation of human C32 melanoma cells caused by inosine (INO) through A_3 adenosine receptor (A_3AR) activation. INO activates, primarily, PLC-PKC-MEK1/2-ERK1/2 pathway after binding to A_3AR, causing cell proliferation. Simultaneously, INO can also enter into the cell through the ENT, which can activates PI3K, contributing also to cell proliferation.

Moreover, we provided proof that A_3AR-MEK1/2-ERK1/2 activation is the major pathway responsible for the C32 proliferative effects elicited by INO. This finding is in agreement with the observation that normal melanocytes do not have detectable ERK activity, requiring growth factors secreted by keratinocytes to survive, whereas melanoma cell lines have enhanced ERK activity due, in part, to autocrine factors stimulation [95].

Taking this into consideration, we propose INO as a stimulation autocrine factor for C32 melanoma cells. Moreover, although the MAPK pathway is primordial for cell proliferation, PI3K pathway is also essential, suggesting that both pathways may be possible targets for future melanoma anticancer therapy. These findings are in agreement with the fact that Ras-Raf-MEK-ERK and PI3K-AKT signalling pathways being constitutively activated in melanoma exerting several key functions in melanoma development and progression [18]. Therefore, these pathways may represent promising therapeutic targets for the effective treatment of metastatic melanoma.

3.3. Potentiation of Cytotoxicity of Paclitaxel in Combination with Cl-IB-MECA in Human C32 Metastatic Melanoma Cells: A New Possible Therapeutic Strategy for Melanoma

At nanomolar concentrations, the selective A_3AR agonist Cl-IB-MECA increased C32 melanoma cell proliferation *via* A_3AR activation [121]. However, Cl-IB-MECA, at micromolar concentrations, caused severe cytotoxicity on C32 cells, through A_3AR-independent mechanisms [121, 130]. These findings are in accordance with other studies reporting that Cl-IB-MECA can inhibit cell proliferation of human thyroid cancer cells, independently of A_3AR activation [59]. Several possibilities could explain this effect, namely, loss of A_3AR selectivity and/or A_3AR desensitization [131]. Other possibilities include i) high concentrations of Cl-IB-MECA can bind to a yet unidentified membrane receptor and trigger downstream signalling or ii) Cl-IB-MECA can be transported into the cell via a NT, and once inside the cell signal through direct interaction with intracellular targets. However, the cytotoxic effect of Cl-IB-MECA in C32 cells was not prevented by the NT inhibitor NBTI [121, 130], discarding the possibility that Cl-IB-MECA enters the cell by a NT-dependent mechanism, in our model. Nevertheless, one cannot discard the last possibility since Cl-IB-MECA can enter the cell through a NT-independent mechanism.

Attempts were also made to explore the potential cytotoxic synergistic effects of the combination of Cl-IB-MECA with a chemotherapeutic agent presently used in therapeutics, paclitaxel (PXT). Combinations of cytotoxic agents may yield a slightly higher survival rate than monotherapies already in use for metastatic melanoma (reviewed in [16]). However, these combinations have been associated with higher toxicities [132]. In addition, melanoma usually relapses because most patients develop resistance to chemotherapeutics, namely to PXT [133]. Our group demonstrated, for the first time, that the cytotoxicity induced by clinically relevant PXT concentrations (low doses) is potentiated by the simultaneous use of Cl-IB-MECA (micromolar concentrations), in human C32 melanoma cells [134]. PXT concentrations used in this study are much lower than those already tested in human bone marrow cells without collateral toxicity [135] and much lower than concentrations found in plasma of PXT-treated cancer patients [136]. A_3AR agonists reveal differential effects on normal and malignant cells, without inhibiting normal cell growth [64] and Cl-IB-MECA is currently being tested in hepatocellular carcinoma patients and considered safe and well-tolerated [62]. Our proposed therapeutic combination seems to act synergistically in causing apoptosis in C32 cells, through enhancement of caspase 9, 8 and 3 activities [134] (Figure 2).

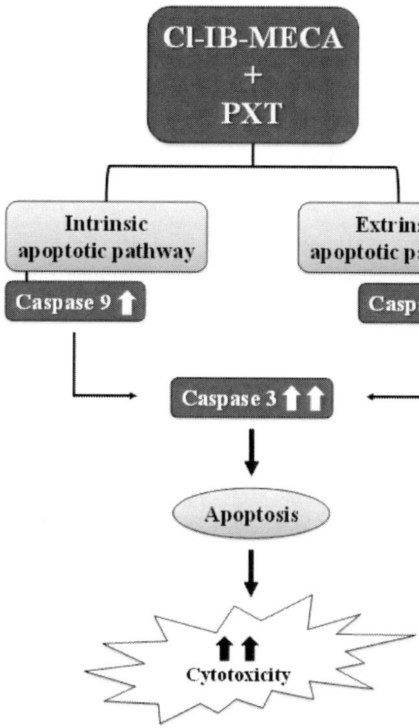

Figure 2. Proposed mechanism for the combinatory therapeutic strategy (Cl-IB-MECA and PXT), on human C32 melanoma cells. Combination of these two drugs enhances activity of intrinsic (caspase 9) and extrinsic pathways (caspase 8) initiator caspase activities, and effector caspase 3 activity, triggering cell apoptosis.

However, the cytotoxicity found for the combination of Cl-IB-MECA and PXT does not involve A_3AR activation. Micromolar concentrations of Cl-IB-MECA were also found to inhibit human thyroid carcinoma cell proliferation by a mechanism independent of the classical A_3AR activation [59]. In fact, when the non-selective AR antagonist (CGS 15943) was combined with Cl-IB-MECA and PTX, cytotoxicity was even more pronounced, suggesting that blocking AR probably prevents the binding of endogenous ADO and/or its metabolic product INO, which could be involved in C32 cell survival. The putative protective role of these molecules is in agreement with previous findings that showed the ability of ADO and/or INO to enhance proliferation of human C32 melanoma cells [121, 124].

3.4. Combination of Cl-IB-MECA with Paclitaxel Is a Highly Effective Cytotoxic Therapy Causing mTOR-Dependent Autophagy and Mitotic Catastrophe on Human Melanoma Cells

The combination of Cl-IB-MECA and low doses PXT revealed a good therapeutic strategy (cytotoxicity) on metastatic melanoma treatment by induction of caspase-dependent apoptosis on human C32 melanoma cells. However, other mechanisms of cell death are definitively involved since the caspase inhibitor (Ac-DEVD-CHO) only prevented approximately 10% of the cytotoxic effects [134]. It is well accepted that the limited success of most current chemotherapeutic agents used in metastatic melanoma is, at least in part, related to cancer cells ability in escaping apoptosis, acquiring drug resistance [20]. Recently, another type of cell death, autophagy, has become an alternative approach to anticancer therapy, as pro-autophagic drugs seem to overcome drug resistance [137].

Efforts were made in order to elucidate the other mechanisms of cell death elicited by the cytotoxic combination of Cl-IB-MECA and PXT, namely the possible involvement of autophagy. Moreover, to increase knowledge on the potential use of this anticancer therapeutic strategy, this combination was also tested in another melanoma cellular model besides C32 cells. Importantly, Cl-IB-MECA in combination with PXT not only causes apoptosis but also induces autophagy, causing cell death on human C32 and A375 melanoma cells [138] (Figure 3). Activation of both apoptosis and autophagy has also been reported to occur in other human melanoma cells treated with a binuclear palladacycle complex [139]. While autophagy represents the main cell death mechanism of the drug combination for C32 cells, in A375 cells apoptosis and autophagy contribute equally to cell death [138]. This discrepancy between C32 and A375 cells concerning their susceptibility to the drug combination in study could be, in part, justified by the gender of each cell line donor, since different metastatic patterns were found on gender of patients with metastatic melanoma and C32 and A375 were collected from male and female donors, respectively [123]. This fact raises again the importance that generalization on gender human melanoma disease should be avoided as the progression of this disease is different in men and women. Therefore, individualized therapies would represent more efficient strategies for melanoma.

Despite mechanisms of autophagy regulation are not yet completely understood, evidence has shown that mTOR (a member of PI3K family) is involved in cell proliferation and plays a major role in preventing autophagy activation [140]. Results obtained with the combination of Cl-IB-MECA and PXT showed that levels of mTOR were decreased in C32 and A375 cells, as revealed by immunocytochemistry. Thus, it is conceivable that this combination inhibits the PI3K pathway, which was found to be constitutively activated in C32 cells [124]. The downregulation of mTOR and autophagy activation was found to be similar to that elicited by the PI3K inhibitor (LY294002) [138]. In fact, autophagy has been reported to be regulated by suppression of mTOR expression [141].

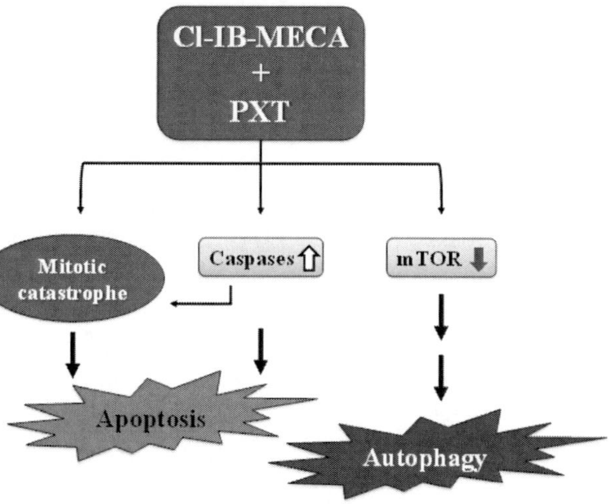

Figure 3. Proposed signalling pathways showing that the combination of Cl-IB-MECA with paclitaxel (PXT) causes different mechanisms of cell death on human C32 and A375 melanoma cell lines.

At present, the PXT mechanisms for cell death are poorly understood [142] and apoptosis was found to be more related to PXT whereas Cl-IB-MECA was with autophagy. Strong disruption of microtubule networks and induction of micro- and multinucleation, hallmarks of mitotic catastrophe, were observed to occur for PXT and for the combination of Cl-IB-MECA with PXT [138]. Moreover, induction of mitotic catastrophe was found to be both caspase-dependent and independent. Although caspase-dependent mitotic catastrophe is being suggested as a special case of apoptosis [143], mitotic catastrophe through mechanisms dependent and/or independent of caspase activation has been also reported [144], corroborating our findings [138]. Overall, these findings provide appealing mechanisms concerning melanoma chemotherapy, as induction of autophagy in melanoma cells that are also undergoing apoptosis, drives more cells to die. Therefore, this therapeutic strategy will potentially improve the effectiveness of PXT through activation of other forms of melanoma cell death.

3.5. The Combination of Cl-IB-MECA with Paclitaxel: A New Anti-Metastatic Therapeutic Strategy for Melanoma

It is well accepted that cellular proliferation is fundamental for melanoma metastasis [145]. The development of metastasis starts with cell migration from the primary tumour, followed by re-adhesion and tissue invasion, with complex interactions between tumour cells and their environment [146]. Based on the variety of cellular mechanisms involved in metastatic progression, attempts were made in order to find new effective drugs that might inhibit tumour cell proliferation, cell migration and/or invasion, and also drugs with anti-angiogenic properties. We had previously demonstrated that INO dramatically increases melanoma cell proliferation [124] and the combination of Cl-IB-MECA with PXT revealed to be very efficient in the induction of melanoma cell death [134, 138]. As such, the metastatic potential of human C32 and A375 melanoma cells was characterized. Afterwards, the role of INO on the promotion of the metastatic mechanisms and ability of the combination of Cl-IB-MECA with PXT in counteracting these effects were also investigated [147]. Our findings were the first demonstration that the combination of Cl-IB-MECA with PXT, is very efficient in preventing the metastatic process, stimulated by INO, on human C32 and A375 melanoma cells by (i) inhibiting cell proliferation; (ii) stopping cell migration; (iii) preventing cell adhesion; (iv) blocking cell invasion through the reduction of

ERK1/2 levels; (v) diminishing colony formation ability; and (vi) reducing angiogenesis (Figure 4).

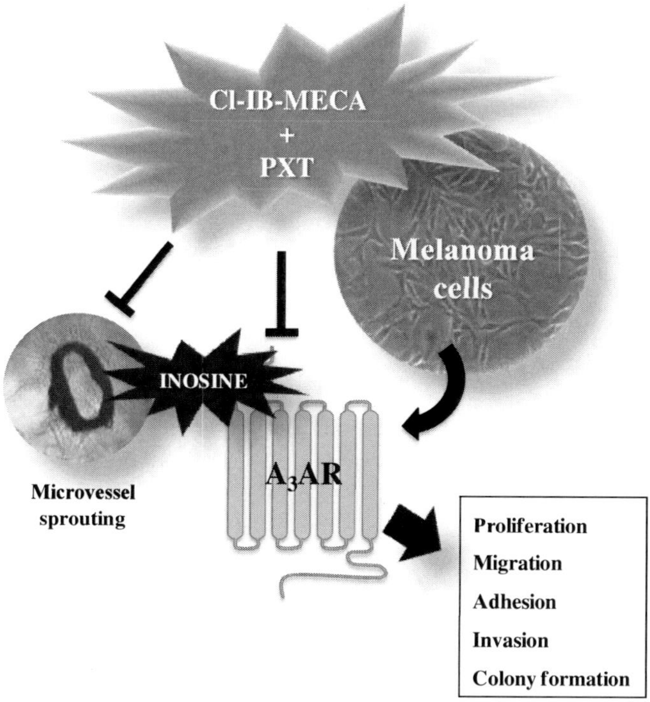

Figure 4. Anti-metastatic activity of the combination of Cl-IB-MECA and PXT on human C32 and A375 melanoma cells, by inhibition of pathways involved in melanoma progression: proliferation, migration, adhesion and invasion, and reduction of ERK1/2 levels, through A_3 adenosine receptor (A_3AR) activation. Inosine was shown to be a stimulating factor of these pathways and increased the ability of A375 cells to form colonies, an effect also mediated by A_3AR activation. Inosine also induced microvessel sprouting and can, therefore, be considered an aggressiveness factor for metastatic progression.

In fact, although INO is considered a weaker but selective A_3AR agonist [148], it works as an aggressiveness agent for these melanoma cells [147]. Moreover, the effects enhanced by INO on human melanoma cells are A_3AR-dependent. Nevertheless, the combination of Cl-IB-MECA with PXT was able to abolish all of them, although with dissimilar importance on the two melanoma cell lines. A surprising finding was the fact that Cl-IB-MECA alone is able to block the metastatic processes enhanced by INO. Although micromolar concentrations of Cl-IB-MECA caused melanoma C32 cell cytotoxicity through AR-independent mechanisms [130, 134], it is conceivable that Cl-IB-MECA may prevent the effects of INO involved in melanoma progression, through interaction with A_3AR. These results are in agreement with works that describe Cl-IB-MECA (at micromolar concentrations) as a silent A_3AR antagonist in the presence of micromolar concentrations of INO [149].

However, it is well established that A_3AR agonists lose their selectivity at high concentrations [150], which suggest that Cl-IB-MECA could also mediate its effects by interaction with other AR subtypes, in these melanoma cellular models. To the best of our knowledge, it is the first time that Cl-IB-MECA is reported to cause anti-migration, anti-adhesion, and anti-invasion effects on human melanoma cells, and this can, at least in part, explain the positive results of this compound in clinical assays [62]. Overall, the combination of Cl-IB-MECA with PXT revealed to be very efficient in stopping metastatic progression, as mechanisms not affected by PXT were impaired upon Cl-IB-MECA addition, thus overcoming the possible PXT chemotherapeutic resistance.

Conclusion

The ability of melanoma to metastasize results in early and general spread, causing very high mortality rates [11]. Resistance to chemotherapy and radiotherapy [16] is the main reason for patients with metastatic melanoma to have an overall survival rate of less than two years [11]. Chemotherapeutic drugs including alkylating agents (dacarbazine), platinum analogues (cisplatin, carboplatin) and anti-microtubular agents (PXT) have been used, alone or in combination, although without significant survival rate improvement [16]. The lack of effective pharmacological approaches could be, at least in part, due to an incomplete understanding of the pathophysiological mechanisms involved in melanoma cells proliferation and, ultimately, tumour progression. Our research led some light on the expression of adenosine receptors in human melanoma cell lines, the role of individual AR subtypes in cell proliferation and cell death, particularly A_3AR, and the underlying mechanisms. Moreover, we showed that the metabolic product of ADO deamination is able to activate A_3AR and stimulate several mechanisms of melanoma progression.

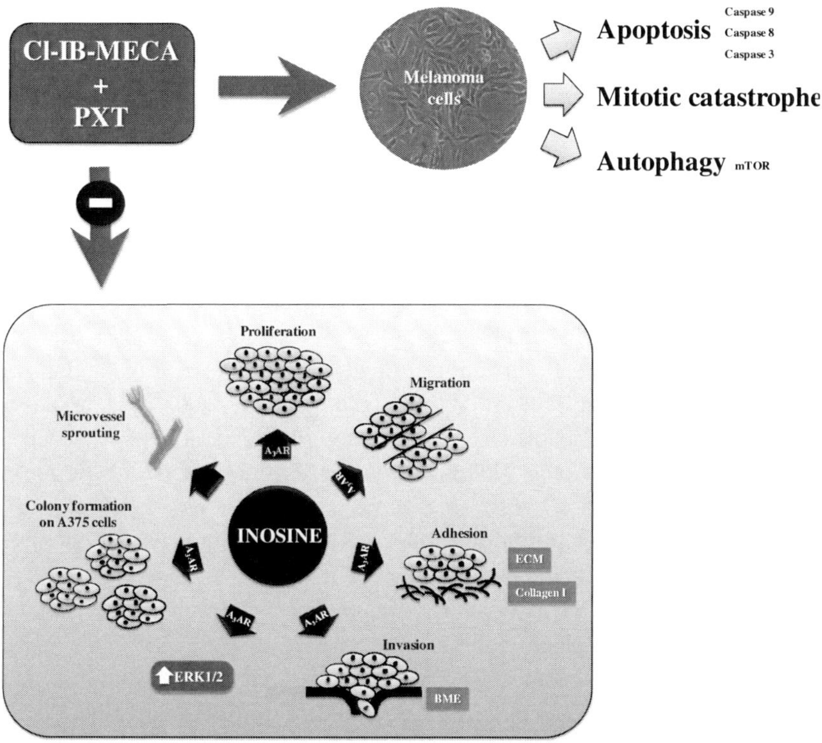

Figure 5. Illustration representing the postulated mechanisms for the cellular signalling pathways altered by the combination of Cl-IB-MECA and PXT. This combination induces multiple mechanisms of cell death (apoptosis with the involvement of mitotic catastrophe, and mTOR-dependent autophagy) and blocks several processes involved in melanoma progression, enhanced by inosine, namely, proliferation, migration, adhesion, invasion, colony formation which are mediated by A_3AR activation, and microvessel sprouting as an angiogenesis indicator.

A particularly important part of our research is related to the proposal of a combinatory approach, using Cl-IB-MECA with low doses PXT, as a new and promising cytotoxic combination for metastatic melanoma treatment. This combination was found to act through the induction of multiple mechanisms of cell death and blockade of several pathways involved in the metastatic potential of melanoma (depicted in Figure 5). This new therapeutic strategy seems to overcome melanoma multiresistance to chemotherapy and to stop, or delay, its progression, therefore, potentially improving metastatic melanoma treatment and, ultimately, patient prognosis.

References

[1] Bergamin LS, Braganhol E, Zanin RF, Edelweiss MI, Battastini AM. Ectonucleotidases in tumor cells and tumor-associated immune cells: an overview. *Journal of biomedicine & biotechnology*. 2012;2012:959848.

[2] Antonioli L, Blandizzi C, Pacher P, Hasko G. Immunity, inflammation and cancer: a leading role for adenosine. *Nature reviews Cancer*. 2013;13(12):842-57.

[3] Burnstock G, Di Virgilio F. Purinergic signalling and cancer. *Purinergic signalling*. 2013;9(4):491-540.

[4] Linden J. Adenosine metabolism and cancer. Focus on "Adenosine downregulates DPPIV on HT-29 colon cancer cells by stimulating protein tyrosine phosphatases and reducing ERK1/2 activity via a novel pathway". *American journal of physiology Cell physiology*. 2006;291(3):C405-6.

[5] Spychala J. Tumor-promoting functions of adenosine. *Pharmacology & therapeutics*. 2000;87(2-3):161-73.

[6] Mahabeleshwar GH, Byzova TV. Angiogenesis in melanoma. *Seminars in oncology*. 2007;34(6):555-65.

[7] Ceruti S, Abbracchio MP. Adenosine signaling in glioma cells. *Advances in experimental medicine and biology*. 2013;986:13-30.

[8] Zhang J, Visser F, King KM, Baldwin SA, Young JD, Cass CE. The role of nucleoside transporters in cancer chemotherapy with nucleoside drugs. *Cancer metastasis reviews*. 2007;26(1):85-110.

[9] Hasko G, Cronstein BN. Adenosine: an endogenous regulator of innate immunity. *Trends in immunology*. 2004;25(1):33-9.

[10] Hasko G, Sitkovsky MV, Szabo C. Immunomodulatory and neuroprotective effects of inosine. *Trends in pharmacological sciences*. 2004;25(3):152-7.

[11] Bandarchi B, Ma L, Navab R, Seth A, Rasty G. From melanocyte to metastatic malignant melanoma. *Dermatology research and practice*. 2010;2010.

[12] Jemal A, Siegel R, Xu J, Ward E. Cancer statistics, 2010. *CA: a cancer journal for clinicians*. 2010;60(5):277-300.

[13] MacKie RM, Hauschild A, Eggermont AM. Epidemiology of invasive cutaneous melanoma. *Annals of oncology : official journal of the European Society for Medical Oncology / ESMO*. 2009;20 Suppl 6:vi1-7.

[14] Eggermont AM, Spatz A, Robert C. Cutaneous melanoma. *Lancet.* 2014;383(9919):816-27.

[15] Erdei E, Torres SM. A new understanding in the epidemiology of melanoma. *Expert review of anticancer therapy.* 2010;10(11):1811-23.

[16] Bhatia S, Tykodi SS, Thompson JA. Treatment of metastatic melanoma: an overview. *Oncology.* 2009;23(6):488-96.

[17] Jones V, Katiyar SK. Emerging phytochemicals for prevention of melanoma invasion. *Cancer letters.* 2013;335(2):251-8.

[18] Meier F, Schittek B, Busch S, Garbe C, Smalley K, Satyamoorthy K, et al. The RAS/RAF/MEK/ERK and PI3K/AKT signaling pathways present molecular targets for the effective treatment of advanced melanoma. *Frontiers in bioscience : a journal and virtual library.* 2005;10:2986-3001.

[19] de Souza CF, Morais AS, Jasiulionis MG. Biomarkers as key contributors in treating malignant melanoma metastases. *Dermatology research and practice.* 2012;2012:156068.

[20] Soengas MS, Lowe SW. Apoptosis and melanoma chemoresistance. *Oncogene.* 2003;22(20):3138-51.

[21] Gorlach A. Control of adenosine transport by hypoxia. *Circulation research.* 2005;97(1):1-3.

[22] Borowiec A, Lechward K, Tkacz-Stachowska K, Skladanowski AC. Adenosine as a metabolic regulator of tissue function: production of adenosine by cytoplasmic 5'-nucleotidases. *Acta biochimica Polonica.* 2006;53(2):269-78.

[23] Linden J. Molecular approach to adenosine receptors: receptor-mediated mechanisms of tissue protection. *Annual review of pharmacology and toxicology.* 2001;41:775-87.

[24] Adair TH. Growth regulation of the vascular system: an emerging role for adenosine. American journal of physiology Regulatory, integrative *and comparative physiology.* 2005;289(2):R283-R96.

[25] Jacobson KA, Gao ZG. Adenosine receptors as therapeutic targets. *Nature reviews Drug discovery.* 2006;5(3):247-64.

[26] Baldwin SA, Beal PR, Yao SY, King AE, Cass CE, Young JD. The equilibrative nucleoside transporter family, SLC29. *Pflugers Archiv : European journal of physiology.* 2004;447(5):735-43.

[27] Eltzschig HK, Abdulla P, Hoffman E, Hamilton KE, Daniels D, Schonfeld C, et al. HIF-1-dependent repression of equilibrative nucleoside transporter (ENT) in hypoxia. *The Journal of experimental medicine.* 2005;202(11):1493-505.

[28] Fredholm BB, AP IJ, Jacobson KA, Klotz KN, Linden J. International Union of Pharmacology. XXV. Nomenclature and classification of adenosine receptors. *Pharmacological reviews*. 2001;53(4):527-52.
[29] Fredholm BB. Adenosine receptors as drug targets. *Experimental cell research*. 2010;316(8):1284-8.
[30] Ballarin M, Fredholm BB, Ambrosio S, Mahy N. Extracellular levels of adenosine and its metabolites in the striatum of awake rats: inhibition of uptake and metabolism. *Acta physiologica Scandinavica*. 1991;142(1):97-103.
[31] Blay J, White TD, Hoskin DW. The extracellular fluid of solid carcinomas contains immunosuppressive concentrations of adenosine. *Cancer research*. 1997;57(13):2602-5.
[32] Ohta A, Gorelik E, Prasad SJ, Ronchese F, Lukashev D, Wong MK, et al. A2A adenosine receptor protects tumors from antitumor T cells. *Proceedings of the National Academy of Sciences of the United States of America*. 2006;103(35):13132-7.
[33] Fishman P, Bar-Yehuda S, Synowitz M, Powell JD, Klotz KN, Gessi S, et al. Adenosine receptors and cancer. *Handbook of experimental pharmacology*. 2009(193):399-441.
[34] Fredholm BB, Irenius E, Kull B, Schulte G. Comparison of the potency of adenosine as an agonist at human adenosine receptors expressed in Chinese hamster ovary cells. *Biochemical pharmacology*. 2001;61(4):443-8.
[35] Schulte G, Fredholm BB. Signalling from adenosine receptors to mitogen-activated protein kinases. *Cellular signalling*. 2003;15(9):813-27.
[36] Gessi S, Merighi S, Varani K, Leung E, Mac Lennan S, Borea PA. The A3 adenosine receptor: an enigmatic player in cell biology. *Pharmacology & therapeutics*. 2008;117(1):123-40.
[37] Zezula J, Freissmuth M. The A(2A)-adenosine receptor: a GPCR with unique features? *British journal of pharmacology*. 2008;153 Suppl 1:S184-90.
[38] Gessi S, Merighi S, Sacchetto V, Simioni C, Borea PA. Adenosine receptors and cancer. *Biochimica et biophysica acta*. 2011;1808(5):1400-12.
[39] Stagg J, Smyth MJ. Extracellular adenosine triphosphate and adenosine in cancer. *Oncogene*. 2010;29(39):5346-58.
[40] Mirza A, Basso A, Black S, Malkowski M, Kwee L, Pachter JA, et al. RNA interference targeting of A1 receptor-overexpressing breast

carcinoma cells leads to diminished rates of cell proliferation and induction of apoptosis. *Cancer biology & therapy*. 2005;4(12):1355-60.

[41] Synowitz M, Glass R, Farber K, Markovic D, Kronenberg G, Herrmann K, et al. A1 adenosine receptors in microglia control glioblastoma-host interaction. *Cancer research*. 2006;66(17):8550-7.

[42] Saito M, Yaguchi T, Yasuda Y, Nakano T, Nishizaki T. Adenosine suppresses CW2 human colonic cancer growth by inducing apoptosis via A(1) adenosine receptors. *Cancer letters*. 2010;290(2):211-5.

[43] Merighi S, Mirandola P, Milani D, Varani K, Gessi S, Klotz KN, et al. Adenosine receptors as mediators of both cell proliferation and cell death of cultured human melanoma cells. *The Journal of investigative dermatology*. 2002;119(4):923-33.

[44] Yasuda Y, Saito M, Yamamura T, Yaguchi T, Nishizaki T. Extracellular adenosine induces apoptosis in Caco-2 human colonic cancer cells by activating caspase-9/-3 via A(2a) adenosine receptors. *Journal of gastroenterology*. 2009;44(1):56-65.

[45] Etique N, Grillier-Vuissoz I, Lecomte J, Flament S. Crosstalk between adenosine receptor (A2A isoform) and ERalpha mediates ethanol action in MCF-7 breast cancer cells. *Oncology reports*. 2009;21(4):977-81.

[46] Montesinos MC, Desai A, Chen JF, Yee H, Schwarzschild MA, Fink JS, et al. Adenosine promotes wound healing and mediates angiogenesis in response to tissue injury via occupancy of A(2A) receptors. *The American journal of pathology*. 2002;160(6):2009-18.

[47] Grant MB, Tarnuzzer RW, Caballero S, Ozeck MJ, Davis MI, Spoerri PE, et al. Adenosine receptor activation induces vascular endothelial growth factor in human retinal endothelial cells. *Circulation research*. 1999;85(8):699-706.

[48] Ryzhov S, Zaynagetdinov R, Goldstein AE, Novitskiy SV, Dikov MM, Blackburn MR, et al. Effect of A2B adenosine receptor gene ablation on proinflammatory adenosine signaling in mast cells. *Journal of immunology*. 2008;180(11):7212-20.

[49] Zeng D, Maa T, Wang U, Feoktistov I, Biaggioni I, Belardinelli L. Expression and function of A2 adenosine receptors in the U87MG tumor cells. *Drug Dev. Res*. 2003;58:405-11.

[50] Merighi S, Benini A, Mirandola P, Gessi S, Varani K, Leung E, et al. Adenosine modulates vascular endothelial growth factor expression via hypoxia-inducible factor-1 in human glioblastoma cells. *Biochemical pharmacology*. 2006;72(1):19-31.

[51] Madi L, Ochaion A, Rath-Wolfson L, Bar-Yehuda S, Erlanger A, Ohana G, et al. The A3 adenosine receptor is highly expressed in tumor versus normal cells: potential target for tumor growth inhibition. *Clinical cancer research : an official journal of the American Association for Cancer Research.* 2004;10(13):4472-9.
[52] Madi L, Bar-Yehuda S, Barer F, Ardon E, Ochaion A, Fishman P. A3 adenosine receptor activation in melanoma cells: association between receptor fate and tumor growth inhibition. *The Journal of biological chemistry.* 2003;278(43):42121-30.
[53] Ohana G, Bar-Yehuda S, Arich A, Madi L, Dreznick Z, Rath-Wolfson L, et al. Inhibition of primary colon carcinoma growth and liver metastasis by the A3 adenosine receptor agonist CF101. *British journal of cancer.* 2003;89(8):1552-8.
[54] Fishman P, Bar-Yehuda S, Ohana G, Barer F, Ochaion A, Erlanger A, et al. An agonist to the A3 adenosine receptor inhibits colon carcinoma growth in mice via modulation of GSK-3 beta and NF-kappa B. *Oncogene.* 2004;23(14):2465-71.
[55] Merighi S, Varani K, Gessi S, Cattabriga E, Iannotta V, Ulouglu C, et al. Pharmacological and biochemical characterization of adenosine receptors in the human malignant melanoma A375 cell line. *British journal of pharmacology.* 2001;134(6):1215-26.
[56] Gessi S, Varani K, Merighi S, Morelli A, Ferrari D, Leung E, et al. Pharmacological and biochemical characterization of A3 adenosine receptors in Jurkat T cells. *British journal of pharmacology.* 2001;134(1):116-26.
[57] Suh BC, Kim TD, Lee JU, Seong JK, Kim KT. Pharmacological characterization of adenosine receptors in PGT-beta mouse pineal gland tumour cells. *British journal of pharmacology.* 2001;134(1):132-42.
[58] Harish A, Hohana G, Fishman P, Arnon O, Bar-Yehuda S. A3 adenosine receptor agonist potentiates natural killer cell activity. *International journal of oncology.* 2003;23(4):1245-9.
[59] Morello S, Petrella A, Festa M, Popolo A, Monaco M, Vuttariello E, et al. Cl-IB-MECA inhibits human thyroid cancer cell proliferation independently of A3 adenosine receptor activation. *Cancer biology & therapy.* 2008;7(2):278-84.
[60] Kim SG, Ravi G, Hoffmann C, Jung YJ, Kim M, Chen A, et al. p53-Independent induction of Fas and apoptosis in leukemic cells by an adenosine derivative, Cl-IB-MECA. *Biochemical pharmacology.* 2002;63(5):871-80.

[61] Morello S, Sorrentino R, Porta A, Forte G, Popolo A, Petrella A, et al. Cl-IB-MECA enhances TRAIL-induced apoptosis via the modulation of NF-kappaB signalling pathway in thyroid cancer cells. *Journal of cellular physiology.* 2009;221(2):378-86.

[62] Stemmer SM, Benjaminov O, Medalia G, Ciuraru NB, Silverman MH, Bar-Yehuda S, et al. CF102 for the treatment of hepatocellular carcinoma: a phase I/II, open-label, dose-escalation study. *The oncologist.* 2013;18(1):25-6.

[63] Montinaro A, Iannone R, Pinto A, Morello S. Adenosine receptors as potential targets in melanoma. *Pharmacological research : the official journal of the Italian Pharmacological Society.* 2013;76:34-40.

[64] Fishman P, Bar-Yehuda S, Madi L, Cohn I. A3 adenosine receptor as a target for cancer therapy. *Anti-cancer drugs.* 2002;13(5):437-43.

[65] Yoshikawa N, Yamada S, Takeuchi C, Kagota S, Shinozuka K, Kunitomo M, et al. Cordycepin (3'-deoxyadenosine) inhibits the growth of B16-BL6 mouse melanoma cells through the stimulation of adenosine A3 receptor followed by glycogen synthase kinase-3beta activation and cyclin D1 suppression. *Naunyn-Schmiedeberg's archives of pharmacology.* 2008;377(4-6):591-5.

[66] Fishman P, Bar-Yehuda S, Barer F, Madi L, Multani AS, Pathak S. The A3 adenosine receptor as a new target for cancer therapy and chemoprotection. *Experimental cell research.* 2001;269(2):230-6.

[67] Merighi S, Benini A, Mirandola P, Gessi S, Varani K, Leung E, et al. A3 adenosine receptors modulate hypoxia-inducible factor-1alpha expression in human A375 melanoma cells. *Neoplasia.* 2005;7(10):894-903.

[68] Merighi S, Simioni C, Gessi S, Varani K, Mirandola P, Tabrizi MA, et al. A(2B) and A(3) adenosine receptors modulate vascular endothelial growth factor and interleukin-8 expression in human melanoma cells treated with etoposide and doxorubicin. *Neoplasia.* 2009;11(10):1064-73.

[69] Chatelain C, Ciarletta P, Ben Amar M. Morphological changes in early melanoma development: influence of nutrients, growth inhibitors and cell-adhesion mechanisms. *Journal of theoretical biology.* 2011;290:46-59.

[70] Liang XS, Pfeiffer RM, Wheeler W, Maeder D, Burdette L, Yeager M, et al. Genetic variants in DNA repair genes and the risk of cutaneous malignant melanoma in melanoma-prone families with/without

CDKN2A mutations. *International journal of cancer Journal international du cancer*. 2012;130(9):2062-6.

[71] Slipicevic A, Herlyn M. Narrowing the knowledge gaps for melanoma. *Upsala journal of medical sciences*. 2012;117(2):237-43.

[72] Bandarchi B, Jabbari CA, Vedadi A, Navab R. Molecular biology of normal melanocytes and melanoma cells. *Journal of clinical pathology*. 2013;66(8):644-8.

[73] Curtin JA, Fridlyand J, Kageshita T, Patel HN, Busam KJ, Kutzner H, et al. Distinct sets of genetic alterations in melanoma. *The New England journal of medicine*. 2005;353(20):2135-47.

[74] Bevona C, Goggins W, Quinn T, Fullerton J, Tsao H. Cutaneous melanomas associated with nevi. *Archives of dermatology*. 2003;139(12):1620-4; discussion 4.

[75] Pfeifer GP, Besaratinia A. UV wavelength-dependent DNA damage and human non-melanoma and melanoma skin cancer. *Photochemical & photobiological sciences : Official journal of the European Photochemistry Association and the European Society for Photobiology*. 2012;11(1):90-7.

[76] Wilkerson BL. Malignant melanoma. *Plastic surgical nursing : official journal of the American Society of Plastic and Reconstructive Surgical Nurses*. 2011;31(3):105-7.

[77] Besaratinia A, Pfeifer GP. Sunlight ultraviolet irradiation and BRAF V600 mutagenesis in human melanoma. *Human mutation*. 2008;29(8):983-91.

[78] Hoerter JD, Bradley P, Casillas A, Chambers D, Weiswasser B, Clements L, et al. Does melanoma begin in a melanocyte stem cell? *Journal of skin cancer*. 2012;2012:571087.

[79] Jhappan C, Noonan FP, Merlino G. Ultraviolet radiation and cutaneous malignant melanoma. *Oncogene*. 2003;22(20):3099-112.

[80] Svobodova A, Walterova D, Vostalova J. Ultraviolet light induced alteration to the skin. *Biomedical papers of the Medical Faculty of the University Palacky*, Olomouc, Czechoslovakia. 2006;150(1):25-38.

[81] McMillan TJ, Leatherman E, Ridley A, Shorrocks J, Tobi SE, Whiteside JR. Cellular effects of long wavelength UV light (UVA) in mammalian cells. *The Journal of pharmacy and pharmacology*. 2008;60(8):969-76.

[82] Li G, Satyamoorthy K, Meier F, Berking C, Bogenrieder T, Herlyn M. Function and regulation of melanoma-stromal fibroblast interactions: when seeds meet soil. *Oncogene*. 2003;22(20):3162-71.

[83] Lee JT, Herlyn M. Microenvironmental influences in melanoma progression. *Journal of cellular biochemistry*. 2007;101(4):862-72.

[84] Haass NK, Smalley KS, Li L, Herlyn M. Adhesion, migration and communication in melanocytes and melanoma. *Pigment cell research / sponsored by the European Society for Pigment Cell Research and the International Pigment Cell Society*. 2005;18(3):150-9.

[85] Gaggioli C, Sahai E. Melanoma invasion - current knowledge and future directions. *Pigment cell research / sponsored by the European Society for Pigment Cell Research and the International Pigment Cell Society*. 2007;20(3):161-72.

[86] Mori S, Chang JT, Andrechek ER, Matsumura N, Baba T, Yao G, et al. Anchorage-independent cell growth signature identifies tumors with metastatic potential. *Oncogene*. 2009;28(31):2796-805.

[87] Michaylira CZ, Nakagawa H. Hypoxic microenvironment as a cradle for melanoma development and progression. *Cancer biology & therapy*. 2006;5(5):476-9.

[88] Bennett DC. How to make a melanoma: what do we know of the primary clonal events? *Pigment cell & melanoma research*. 2008;21(1):27-38.

[89] Meier F, Busch S, Lasithiotakis K, Kulms D, Garbe C, Maczey E, et al. Combined targeting of MAPK and AKT signalling pathways is a promising strategy for melanoma treatment. *The British journal of dermatology*. 2007;156(6):1204-13.

[90] Hilger RA, Scheulen ME, Strumberg D. The Ras-Raf-MEK-ERK pathway in the treatment of cancer. *Onkologie*. 2002;25(6):511-8.

[91] Rodolfo M, Daniotti M, Vallacchi V. Genetic progression of metastatic melanoma. *Cancer letters*. 2004;214(2):133-47.

[92] Lazar-Molnar E, Hegyesi H, Toth S, Falus A. Autocrine and paracrine regulation by cytokines and growth factors in melanoma. *Cytokine*. 2000;12(6):547-54.

[93] McCubrey JA, Steelman LS, Chappell WH, Abrams SL, Wong EW, Chang F, et al. Roles of the Raf/MEK/ERK pathway in cell growth, malignant transformation and drug resistance. *Biochimica et biophysica acta*. 2007;1773(8):1263-84.

[94] Godeny MD, Sayeski PP. ERK1/2 regulates ANG II-dependent cell proliferation via cytoplasmic activation of RSK2 and nuclear activation of elk1. *American journal of physiology Cell physiology*. 2006;291(6):C1308-17.

[95] Smalley KS. A pivotal role for ERK in the oncogenic behaviour of malignant melanoma? *International journal of cancer Journal international du cancer*. 2003;104(5):527-32.
[96] Kortylewski M, Heinrich PC, Kauffmann ME, Bohm M, MacKiewicz A, Behrmann I. Mitogen-activated protein kinases control p27/Kip1 expression and growth of human melanoma cells. *The Biochemical journal*. 2001;357(Pt 1):297-303.
[97] Vivanco I, Sawyers CL. The phosphatidylinositol 3-Kinase AKT pathway in human cancer. *Nature reviews Cancer*. 2002;2(7):489-501.
[98] Curtin JA, Stark MS, Pinkel D, Hayward NK, Bastian BC. PI3-kinase subunits are infrequent somatic targets in melanoma. *The Journal of investigative dermatology*. 2006;126(7):1660-3.
[99] Singh RS, Diwan AH, Zhang PS, Prieto VG. Phosphoinositide 3-kinase is not overexpressed in melanocytic lesions. *Journal of cutaneous pathology*. 2007;34(3):220-5.
[100] Robertson GP. Functional and therapeutic significance of Akt deregulation in malignant melanoma. *Cancer metastasis reviews*. 2005;24(2):273-85.
[101] Dhawan P, Singh AB, Ellis DL, Richmond A. Constitutive activation of Akt/protein kinase B in melanoma leads to up-regulation of nuclear factor-kappaB and tumor progression. *Cancer research*. 2002;62(24):7335-42.
[102] Dai DL, Martinka M, Li G. Prognostic significance of activated Akt expression in melanoma: a clinicopathologic study of 292 cases. *Journal of clinical oncology : official journal of the American Society of Clinical Oncology*. 2005;23(7):1473-82.
[103] Velho TR. Metastatic melanoma - a review of current and future drugs. *Drugs in context*. 2012;2012:212242.
[104] Serrone L, Zeuli M, Sega FM, Cognetti F. Dacarbazine-based chemotherapy for metastatic melanoma: thirty-year experience overview. *Journal of experimental & clinical cancer research* : CR. 2000;19(1):21-34.
[105] Chiarion Sileni V, Nortilli R, Aversa SM, Paccagnella A, Medici M, Corti L, et al. Phase II randomized study of dacarbazine, carmustine, cisplatin and tamoxifen versus dacarbazine alone in advanced melanoma patients. *Melanoma research*. 2001;11(2):189-96.
[106] Wiernik PH, Einzig AI. Taxol in malignant melanoma. *Journal of the National Cancer Institute Monographs*. 1993(15):185-7.

[107] Nathan FE, Berd D, Sato T, Mastrangelo MJ. Paclitaxel and tamoxifen: An active regimen for patients with metastatic melanoma. *Cancer.* 2000;88(1):79-87.
[108] Flaherty KT. New molecular targets in melanoma. *Current opinion in oncology.* 2004;16(2):150-4.
[109] Atkins MB, Lotze MT, Dutcher JP, Fisher RI, Weiss G, Margolin K, et al. High-dose recombinant interleukin 2 therapy for patients with metastatic melanoma: analysis of 270 patients treated between 1985 and 1993. *Journal of clinical oncology : official journal of the American Society of Clinical Oncology.* 1999;17(7):2105-16.
[110] Atkins MB, Kunkel L, Sznol M, Rosenberg SA. High-dose recombinant interleukin-2 therapy in patients with metastatic melanoma: long-term survival update. *The cancer journal from Scientific American.* 2000;6 Suppl 1:S11-4.
[111] Hodi FS, O'Day SJ, McDermott DF, Weber RW, Sosman JA, Haanen JB, et al. Improved survival with ipilimumab in patients with metastatic melanoma. *The New England journal of medicine.* 2010;363(8):711-23.
[112] Finn L, Markovic SN, Joseph RW. Therapy for metastatic melanoma: the past, present, and future. *BMC medicine.* 2012;10:23.
[113] Eggermont AM, Robert C. New drugs in melanoma: it's a whole new world. *European journal of cancer.* 2011;47(14):2150-7.
[114] Chapman PB, Hauschild A, Robert C, Haanen JB, Ascierto P, Larkin J, et al. Improved survival with vemurafenib in melanoma with BRAF V600E mutation. *The New England journal of medicine.* 2011;364(26):2507-16.
[115] Hauschild A, Grob JJ, Demidov LV, Jouary T, Gutzmer R, Millward M, et al. Dabrafenib in BRAF-mutated metastatic melanoma: a multicentre, open-label, phase 3 randomised controlled trial. *Lancet.* 2012;380(9839):358-65.
[116] Heidorn SJ, Milagre C, Whittaker S, Nourry A, Niculescu-Duvas I, Dhomen N, et al. Kinase-dead BRAF and oncogenic RAS cooperate to drive tumor progression through CRAF. *Cell.* 2010;140(2):209-21.
[117] Flaherty KT, Infante JR, Daud A, Gonzalez R, Kefford RF, Sosman J, et al. Combined BRAF and MEK inhibition in melanoma with BRAF V600 mutations. *The New England journal of medicine.* 2012;367(18):1694-703.
[118] End DW, Smets G, Todd AV, Applegate TL, Fuery CJ, Angibaud P, et al. Characterization of the antitumor effects of the selective farnesyl

protein transferase inhibitor R115777 in vivo and in vitro. *Cancer research.* 2001;61(1):131-7.

[119] Margolin KA, Moon J, Flaherty LE, Lao CD, Akerley WL, 3rd, Othus M, et al. Randomized phase II trial of sorafenib with temsirolimus or tipifarnib in untreated metastatic melanoma (S0438). *Clinical cancer research : an official journal of the American Association for Cancer Research.* 2012;18(4):1129-37.

[120] Lopez-Fauqued M, Gil R, Grueso J, Hernandez-Losa J, Pujol A, Moline T, et al. The dual PI3K/mTOR inhibitor PI-103 promotes immunosuppression, in vivo tumor growth and increases survival of sorafenib-treated melanoma cells. *International journal of cancer Journal international du cancer.* 2010;126(7):1549-61.

[121] Falcato-Soares A, Diniz C, Fresco P. Characterization of adenosine receptors in human and mouse melanoma cells. *Medimond International Proceedings: 6th European Congress of Pharmacology, Granada (Spain).* 2013:29-34.

[122] Panjehpour M, Castro M, Klotz KN. Human breast cancer cell line MDA-MB-231 expresses endogenous A2B adenosine receptors mediating a Ca2+ signal. *British journal of pharmacology.* 2005;145(2):211-8.

[123] Mervic L. Time course and pattern of metastasis of cutaneous melanoma differ between men and women. *PloS one.* 2012;7(3):e32955.

[124] Soares AS, Costa VM, Diniz C, Fresco P. Inosine Strongly Enhances Proliferation of Human C32 Melanoma Cells through PLC-PKC-MEK1/2-ERK1/2 and PI3K Pathways. *Basic & clinical pharmacology & toxicology.* 2014.

[125] Bavaresco L, Bernardi A, Braganhol E, Cappellari AR, Rockenbach L, Farias PF, et al. The role of ecto-5'-nucleotidase/CD73 in glioma cell line proliferation. *Molecular and cellular biochemistry.* 2008;319(1-2):61-8.

[126] Mazurek S, Michel A, Eigenbrodt E. Effect of extracellular AMP on cell proliferation and metabolism of breast cancer cell lines with high and low glycolytic rates. *The Journal of biological chemistry.* 1997;272(8):4941-52.

[127] Pirincci N, Gecit I, Gunes M, Yuksel MB, Kaba M, Tanik S, et al. Serum adenosine deaminase, catalase and carbonic anhydrase activities in patients with bladder cancer. *Clinics.* 2012;67(12):1443-6.

[128] Aghaei M, Karami-Tehrani F, Salami S, Atri M. Adenosine deaminase activity in the serum and malignant tumors of breast cancer: the

assessment of isoenzyme ADA1 and ADA2 activities. *Clinical biochemistry.* 2005;38(10):887-91.

[129] Zanini D, Schmatz R, Pelinson LP, Pimentel VC, da Costa P, Cardoso AM, et al. Ectoenzymes and cholinesterase activity and biomarkers of oxidative stress in patients with lung cancer. *Molecular and cellular biochemistry.* 2013;374(1-2):137-48.

[130] Soares AF, Diniz C, Fresco P. A3 adenosine receptor effects on malignant melanoma cells. *FEBS J.* 2012;279(Suppl. 1):547.

[131] Klaasse EC, Ijzerman AP, de Grip WJ, Beukers MW. Internalization and desensitization of adenosine receptors. *Purinergic signalling.* 2008;4(1):21-37.

[132] Mouawad R, Sebert M, Michels J, Bloch J, Spano JP, Khayat D. Treatment for metastatic malignant melanoma: old drugs and new strategies. *Critical reviews in oncology/hematology.* 2010;74(1):27-39.

[133] van Herpen CM, Eskens FA, de Jonge M, Desar I, Hooftman L, Bone EA, et al. A Phase Ib dose-escalation study to evaluate safety and tolerability of the addition of the aminopeptidase inhibitor tosedostat (CHR-2797) to paclitaxel in patients with advanced solid tumours. *British journal of cancer.* 2010;103(9):1362-8.

[134] Soares AS, Costa VM, Diniz C, Fresco P. Potentiation of cytotoxicity of paclitaxel in combination with Cl-IB-MECA in human C32 metastatic melanoma cells: A new possible therapeutic strategy for melanoma. *Biomedicine & pharmacotherapy = Biomedecine & pharmacotherapie.* 2013;67(8):777-89.

[135] Pessina A, Bonomi A, Cocce V, Invernici G, Navone S, Cavicchini L, et al. Mesenchymal stromal cells primed with paclitaxel provide a new approach for cancer therapy. *PloS one.* 2011;6(12):e28321.

[136] Bulitta JB, Zhao P, Arnold RD, Kessler DR, Daifuku R, Pratt J, et al. Mechanistic population pharmacokinetics of total and unbound paclitaxel for a new nanodroplet formulation versus Taxol in cancer patients. *Cancer chemotherapy and pharmacology.* 2009;63(6):1049-63.

[137] Hanahan D, Weinberg RA. Hallmarks of cancer: the next generation. *Cell.* 2011;144(5):646-74.

[138] Soares AS, Costa VM, Diniz C, Fresco P. Combination of ClIBMECA with paclitaxel is a highly effective cytotoxic therapy causing mTORdependent autophagy and mitotic catastrophe on human melanoma cells. *Journal of cancer research and clinical oncology.* 2014;140(6):921-35.

[139] Aliwaini S, Swarts AJ, Blanckenberg A, Mapolie S, Prince S. A novel binuclear palladacycle complex inhibits melanoma growth in vitro and in vivo through apoptosis and autophagy. *Biochemical pharmacology.* 2013;86(12):1650-63.

[140] Mizushima N. Autophagy: process and function. *Genes & development.* 2007;21(22):2861-73.

[141] Quidville V, Alsafadi S, Goubar A, Commo F, Scott V, Pioche-Durieu C, et al. Targeting the deregulated spliceosome core machinery in cancer cells triggers mTOR blockade and autophagy. *Cancer research.* 2013;73(7):2247-58.

[142] Bhalla KN. Microtubule-targeted anticancer agents and apoptosis. *Oncogene.* 2003;22(56):9075-86.

[143] Castedo M, Perfettini JL, Roumier T, Valent A, Raslova H, Yakushijin K, et al. Mitotic catastrophe constitutes a special case of apoptosis whose suppression entails aneuploidy. *Oncogene.* 2004;23(25):4362-70.

[144] Mansilla S, Priebe W, Portugal J. Mitotic catastrophe results in cell death by caspase-dependent and caspase-independent mechanisms. *Cell cycle.* 2006;5(1):53-60.

[145] Madonna G, Ullman CD, Gentilcore G, Palmieri G, Ascierto PA. NF-kappaB as potential target in the treatment of melanoma. *Journal of translational medicine.* 2012;10:53.

[146] Attoub S, Arafat K, Gelaude A, Al Sultan MA, Bracke M, Collin P, et al. Frondoside a suppressive effects on lung cancer survival, tumor growth, angiogenesis, invasion, and metastasis. *PloS one.* 2013;8(1):e53087.

[147] Soares AS, Costa VM, Diniz C, Fresco P. The combination of Cl-IB-MECA with paclitaxel: a new anti-metastatic therapeutic strategy for melanoma. *Cancer chemotherapy and pharmacology*. 2014.

[148] Jin X, Shepherd RK, Duling BR, Linden J. Inosine binds to A3 adenosine receptors and stimulates mast cell degranulation. *The Journal of clinical investigation.* 1997;100(11):2849-57.

[149] Wolber C, Fozard JR. The receptor mechanism mediating the contractile response to adenosine on lung parenchymal strips from actively sensitised, allergen-challenged Brown Norway rats. *Naunyn-Schmiedeberg's archives of pharmacology*. 2005;371(2):158-68.

[150] Mlejnek P, Dolezel P, Frydrych I. Effects of synthetic A3 adenosine receptor agonists on cell proliferation and viability are receptor independent at micromolar concentrations. *Journal of physiology and biochemistry*. 2013;69(3):405-17.

In: Adenosine Receptors
Editor: Kasandra Warrick
ISBN: 978-1-63463-454-0

Chapter IV

Urinary Bladder Disorders: Is Adenosine Friend or Foe?

M. Silva-Ramos[1,2,3,#,*], I. Silva[1,2,#], M. Faria[1,2], M. T. Magalhães-Cardoso[1,2] and P. Correia-de-Sá[1,2*]

[1]Laboratório de Farmacologia e Neurobiologia/UMIB,
[2]Center for Drug Discovery and Innovative Medicines (MedInUP),
Instituto de Ciências Biomédicas de Abel Salazar (ICBAS),
Universidade do Porto, and [3]Serviço de Urologia,
Centro Hospitalar do Porto (CHP), Porto, Portugal

Abstract

There is a considerable bulk of evidence for the relevance of ATP in the regulation of bladder function in health and in disease conditions. However the actions of its metabolite, adenosine, on bladder tissue only recently are being unveiled. All four subtypes of adenosine receptors (A_1, $A_{2a}A_{2A}$, A_{2B} and A_3) have been identified in the bladder of experimental

* Corresponding author: Paulo Correia-de-Sá, MD, PhD, Laboratório de Farmacologia e Neurobiologia, Unit for Multidisciplinary Investigations in Biomedicine (UMIB) and Center for Drug Discovery and Innovative Medicines (MedInUP), Instituto de Ciências Biomédicas de Abel Salazar (ICBAS) - Universidade do Porto (UP), R. Jorge Viterbo Ferreira, 228, 4050-313 Porto, Portugal. Tel.: 351-220428212; fax: 351-220428090. *E-mail address:* farmacol@icbas.up.pt.

# These authors contributed equally to this work.

animals by RT-PCR and Western blot analysis; these receptors are differently distributed among the urothelium and detrusor layers. Adenosine receptors are also present in the human bladder where they can modulate filling sensations and voiding. While adenosine inhibits nerve-evoked acetylcholine release from cholinergic nerve efferents, via high affinity A_1 receptors, high concentrations of the nucleoside are required to reduce detrusor contractions induced by carbachol, ATP and potassium depolarizations. Characterization of the adenosine receptor involved in detrusor smooth muscle relaxation is still a matter of debate. Besides the major effects of the nucleoside on bladder voiding, adenosine also influences the filling phase of the micturition reflex by controlling the release of ATP from urothelial cells and the firing of sensory nerve afferents, via A_1 receptors activation. Differences in the kinetics of ATP metabolism by ectonucleotidases and adenosine biosynthesis between luminal and serosal sides of the urothelium may explain the dominant role of adenosine in the suburothelium and detrusor smooth muscle layers. Thus, the adenosine signaling pathway is now being explored as a potential therapeutic target for human bladder disorders, including inflammatory diseases, overactivity and outflow obstruction.

Keywords: Adenosine receptors, ATP release, Acetylcholine release, Ectonucleotidases, Human urinary bladder, Lower urinary tract symptoms.

Introduction

The lower urinary tract, comprising bladder and outflow tract, has two main functions: urine storage and voiding of the bladder content at most convenient times. The urinary bladder functions as a low-pressure reservoir for most of the time: the filling phase. During this phase the intravesical pressure is low and the outflow resistance high. During voiding the reverse occurs, that is intravesical pressure rises and the outflow pathway relaxes to begin initiate voiding. The initiation and control of these changes of function are carried out by a complex interplay between sensory mechanisms within the bladder, coordinate responses commanded by the brain and sacral spinal cord, and motor control over bladder and outflow muscle tone (Birder et al., 2010). This process is not simple, thus explaining why for some any reason around 60% of adults' present lower urinary tract symptoms (Irwin et al., 2006).

It is now accepted that purines, namely ATP, are involved in a number of physiological processes in the lower urinary tract. ATP was shown (1) to be a co-transmitter with acetylcholine (ACh) in the parasympathetic control of bladder contraction (Burnstock et al., 1972), (2) to activate sensory nerves during bladder filling conveying both normal and abnormal sensations like urgency and pain (Birder and Andersson, 2013; Ferguson et al., 1997; Wang et al., 2005), and (3) to participate in the central control of bladder reflexes (Rocha et al., 2001). However, not so much is known on the effects of the ATP metabolite adenosine. First reports of the effect of adenosine in the lower urinary tract are from the nineteen seventies last century and coincided with Burnstock's group attempt to prove that ATP was the non-cholinergic non-adrenergic neurotransmitter on nerves supplying the urinary bladder (Burnstock et al., 1978b; Burnstock et al., 1972). They managed to show that the breakdown product of ATP, adenosine, caused a decrease in both the tone and the spontaneous activity of the urinary bladder (Burnstock et al., 1978a).

Adenosine is a ubiquitously produced nucleoside that participates in the normal function of the cardiovascular, respiratory, renal, gastrointestinal, neuronal, musculoskeletal and immune systems where it regulates diverse phenomena including neuronal excitability, vasodilation, smooth muscle relaxation, protection against ischemic and inflammatory insults, and the release of vasoactive and neuruoactive substances. Adenosine is generated within cells from the hydrolysis of *S*-adenosyl-L-homocysteine in parallel to its formation from catabolism of adenine nucleotides, ATP, ADP, AMP or cAMP, via nucleotidases, which existing both at the intracellular and the extracellular space (Fredholm et al., 2001; Jackson and Raghvendra, 2004). Once produced, adenosine may undergo bidirectional translocation between cytoplasm and the extracellular fluid (through equilibrative or concentrative nucleoside transporters), deamination into inosine by adenosine deaminase (ADA, EC 3.5.4.4), and phosphorylation by intracellular adenosine kinase (AK, EC 2.7.1.20) to generate AMP. The way adenosine builds its influence to control cells communication depends on the extracellular concentration of the nucleoside, which is achieved by balancing extracellular adenosine generation and inactivation mechanisms (cellular uptake and/or extracellular deamination) (Correia-de-Sá and Ribeiro, 1996; Gonçcalves and Queiroz, 1993). Although adenosine transport may be faster than its subsequent intracellular metabolism (for a review see, Geiger, 1991), numerous studies have shown that phosphorylation of adenosine by adenosine kinase is the primary metabolic pathway regulating both intra- and extracellular levels of the nucleoside (Arch and Newsholme, 1978; Lloyd and Fredholm, 1995). In the urinary bladder, parallel formation of adenosine from the hydrolysis of ATP released from stimulated neuronal and non-neuronal cells (e.g. urothelial cells, interstitial cells, fibroblast-like cells, smooth muscle fibers) is also possible. Therefore, it is conceivable that adenosine contributes to an overall homeostatic role on bladder activity during instability conditions (*e.g.* inflammation, hypoxic/ischemic insults, outlet obstruction) when endogenous adenosine becomes elevated. Extracellular inactivation may restrict adenosine actions to the release/production sites and may limit diffusion of the exogenously added nucleoside towards the active zones (see e.g. Correia-de-Sá et al., 2006; Duarte-Araujo et al., 2004).

Once in the extracellular milieu, it adenosine may interact with membrane-bound P1 receptors (Fredholm et al., 2001). Four subtypes of adenosine receptors have been defined, A_1, A_{2A}, A_{2B} and A_3, all of which are G protein-coupled receptors with seven transmembrane domains. A_1 and A_3 receptors preferably interact with members of the G_i family and inactive adenylate cyclase to decrease the production of cAMP, whereas A_{2A} and A_{2B} receptors are coupled to G_s and stimulate cAMP production. All four receptors activate phospholipase C (PLC), resulting in inositol 3,4,5-trisphosphate (IP_3) production and increased cytoplasmic Ca^{2+} (Klinger et al., 2002), and can also stimulate mitogen-activated protein kinase cascades (Schulte and Fredholm, 2003).

In humans, the A_3, A_1 and A_{2A} receptor subtypes are activated by submicromolar concentrations of adenosine (0.29 μM, 0.31 μM, and 0.73 μM, respectively) while the adenosine A_{2B} receptor is activated at much higher concentrations of the nucleoside (23.5 μM) (Fredholm et al., 2001). This pattern differs significantly from that found in other mammals where the A_3 receptor has low affinity for adenosine. These findings suggest that the human adenosine A_1, A_{2A} and A_3 subtypes respond to physiological concentrations of adenosine (between 30 and 300 nM), while the A_{2B} receptor subtype requires higher concentrations of the agonist for activation which may rise from 10 to 1000 fold in pathological conditions.

Adenosine Receptors Expression in the Urinary Bladder

First studies aiming at characterizing pharmacologically P1 receptors in the urinary bladder were performed in mice and supported the role of both pre-synaptic (via A_1 receptors) and post-synaptic (via A_2 receptors) effects (Acevedo et al., 1992; Nicholls et al., 1992). The role of low affinity A_{2B} and A_3 receptor subtypes has been more difficult to characterize (see below).

Paradoxically, RT-PCR analysis has detected all four adenosine receptor subtypes in the rat urinary bladder: A_{2A} and A_{2B} being more expressed than A_1, and whereas A_3 being was the least expressed (Dixon et al., 1996). A similar pattern was detected for the mRNA expression of adenosine receptors in the human bladder epithelial carcinoma T24 cell line (Phelps et al., 2006). Another study in female rats confirmed the predominance of the A_{2B} receptor mRNA in the urinary bladder and reported increases in P1 receptor mRNA with age (Owen et al., 2012). Because PCR results do not necessarily correlate with the presence of functional adenosine receptors displayed on the cell surface, further characterization of these data by Western blot analysis and functional assays are highly advisablerecommended.

Using Western blot analysis and immunofluorescence localization, Yu and col. (2006) detected the expression of all four types of adenosine receptors in the bladder urothelium of female Sprague-Dawley rats (Yu et al., 2006). The A_1 receptors were predominantly localized to the apical membrane of the umbrella cell layers, whereas A_{2A}, A_{2B} and A_3 receptors were localized intracellularly or on the basolateral membrane of umbrella cells and the plasma membrane of underlying cell layers. Signals for the A_1 receptor were also observed in the underlying submucosal connective tissue, but suburothelial tissue elements (possibly connective cells, myofibroblasts or blood vessels) were more strongly labeled with the A_{2A} antibody. Although information regarding location of adenosine receptors on the detrusor is still scant, there are reports suggesting that A_{2B} receptors are abundantly expressed in rat bladder smooth muscle (Stehle et al., 1992; Yu et al., 2006). In addition, there are pharmacological evidences for the presence of A_1, A_{2A} and A_3 receptors on detrusor muscle, often with conflicting results depending on the species and experimental settings (Gopalakrishnan et al., 2002; Vesela et al., 2011; Yang et al., 2000). Surprisingly, there is little data characterizing the adenosine receptor subtype(s) involved in the human urinary bladder function.

Effects of Adenosine in the Urothelium

The uroepithelium or urothelium is a transition epithelium comprised of three cell layers, named from top to bottom umbrella cells, intermediate and basal layers. The urothelium forms a relatively impermeable barrier that is maintained in part by regulated membrane turnover in the outermost umbrella cells. The uroepithelium has a newly recognized sensory function which detects bladder fullness signaling to the underlying tissues (Apodaca, 2004; Birder, 2005). Between the urothelium and the detrusor smooth muscle layer is the suburothelium, which contains a neuronal network, a functional synciytium of interstitial cells, blood vessels, and some muscle fibers at the inner face. Upon exposure to external stresses the urothelium releases a number of signal molecules (e.g. nitric oxide, acetylcholine, ATP) that might directly influence afferent suburothelial nerve activity or diffuse deeplypth into the detrusor layer to alter smooth muscle contractility (Birder et al., 2012; Sui et al., 2008). Besides this most studied paracrine action, signaling molecules may also exert autocrine functions influencing urothelial cells activity.

As the bladder fills, the apical surface of umbrella cells becomes flattened and cytoplasmic discoidal/fusiform vesicles fuse with the apical plasma membrane, which increases the luminal surface area of the bladder and may allow the bladder to accommodate increased volume of urine (Truschel et al., 2002). Beyond the role of uroepithelium in the bladder filling, it is also associated with mechanotransduction pathways that increase cyclic AMP and Ca^{2+} within the uroepithelium (Wang et al., 2003). Studies indicate that the uroepithelium is intimately associated with the sensory nervous system. Capsaicin-sensitive transient receptor potential channel, vanilloid subfamily member 1 (TRPV1)-positive primary afferent neurons are localized near to and within the uroepithelium, and are also in close proximity to blood vessels and smooth muscle cells (Birder and de Groat, 2007; Maggi, 1990). In addition, cholinergic (choline acetyltransferase positive) nerve fibers are detected just below the uroepithelium, which may indicate the presence of efferent innervation as well (Gillespie et al., 2006). In summary, to act as the primary barrier in the bladder, data indicate that the uroepithelium may function as an integral part of a 'sensory web'. This web includes the uroepithelium, closely apposed nerve fibers, interstitial cells, including myofibroblasts and mast cells, and the detrusor/sphincter muscles. The overall purpose of the sensory web is to coordinate the function of the bladder and its associated tissues to changes in the bladder content (Apodaca et al., 2007).

There are innumerous studies demonstrating that the urothelium can release ATP in response to stretch and chemical stimuli, in animals and in humans (Ferguson et al., 1997; Timoteo et al., 2014; Wang et al., 2003), but it is unknown whether this contributes to the high concentrations of adenosine found in urine (Heyne et al., 2004; Vidotto et al., 2003). ATP measured in the urine is significantly higher in women with overactive bladder; this led our group to consider that urinary ATP may be a dynamic biomarker of detrusor activity in women with overactive bladder syndrome (Silva-Ramos et al., 2013). Once released from urothelium, ATP is rapidly dephosphorylated by four main families of ectonucleotidases, both soluble and membrane-bound: NTPDases (ectonucleoside triphosphate diphosphohydrolases), NPPs (nucleotide pyrophosphatase/phos- phodiesterases), alkaline phosphatases and ecto-5'-nucleotidase (Robson et al., 2006). NTPDase1 (CD39 or apyrase) dephosphorylates ATP directly into AMP, with minimal accumulation of ADP. NTPDase2 (ATPase) is a preferential nucleoside triphosphatase hydrolysing ADP 10 to 15 times less efficiently than ATP, leading to minimal AMP accumulation (Kukulski et al., 2005). NTPDase3 and NTPDase8 are functional intermediates between NTPDase1 and NTPDase2 (Kukulski et al., 2005). Because of their involvement in physiological processes, namely blood clotting, vascular inflammation, immune reactions and certain types of cancer, NTPDases are now considered as potential drug targets (Gendron et al., 2002). As for NPP1 and NPP3, they release nucleoside 5'-monophosphate from a variety of nucleotides, but intriguingly, their phosphorylated product (e.g. AMP) bind to NPPs with a higher affinity than substrates do, and thus inhibit catalysis (Stefan et al., 2005). The ecto-5'-nucleotidases/CD73 is capable of forming adenosine and inorganic phosphate from the hydrolysis of AMP (Colgan et al., 2006). Finally, the alkaline phosphatases are a family of abundant enzymes with broad range specificity, including dephosphorylation of ATP, ADP as well as AMP. The overall role of this family in purinergic signaling is, however, not that well investigated (Zimmermann, 2000).

It has been demonstrated that about 50% of the extracellular adenosine available is the result of the break-down of ATP (Smith and Lu, 1991) which is currently regarded to be the most important route of generating extracellular adenosine (Zhang and Xia, 2012). *In vitro* degradation studies of ATP in a guinea-pig detrusor strip organ bath setup showed that ATP was rapidly broken down within minutes to form mainly, but not exclusively, adenosine (Cusack and Hourani, 1984). Magnesium-dependent adenosine triphosphatase (Mg^{2+}-ATPase) as well as 5′-nucleotidase and alkaline phosphatase were identified in epithelial cells of the rat urinary bladder (Zhang et al., 1991). The four members of the NTPDase family, as well as 5′-nucleotidase, are expressed in the mouse bladder, as demonstrated by RT-PCR (Yu et al., 2011). However, the topology of these ecto-enzymes differs among the various layers of the bladder wall. Immunofluorescence studies failed to identify NTPDase1 and NTPDase2 in the urothelium, only NTPDase3 showed some labeling in the intermediate and basal cells and NTPDase8 revealed diffuse labeling throughout the epithelium; 5′-Nucleotidase was present exclusively in detrusor smooth muscle together with NTPDase1. These results differ substantially from our findings in the human urothelium, where NTPDase1, NTPDase2 and NTPDase3 antibodies label the apical plasma membrane of umbrella cells, whereas the intermediate layer stained positively with NTPDase1 and NTPDase2, but no NTPDase3 immunoreactivity was found (Correia-de-Sá, 2010). The most expressed NTPDase2 is the main contributor to the extracellular catabolism of ATP leading to a preferential ADP accumulation with low adenosine formation at the luminal surface of the isolated human urothelium.

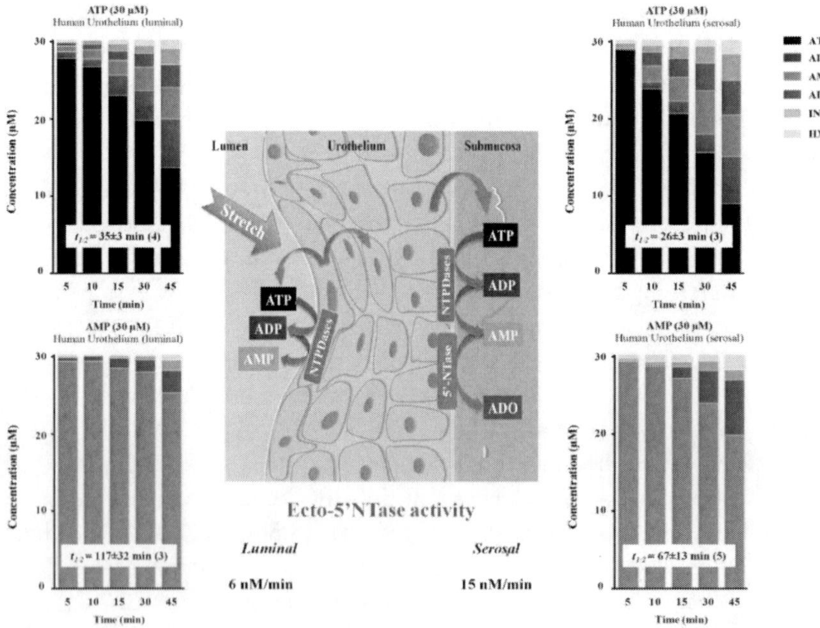

Figure 1. The catabolism of adenine nucleotides and adenosine formation by ecto-5'-nucleotidase is faster in the serosal than in the luminal side of the human urothelium. Illustrated is the time course of the extracellular catabolism of adenine nucleotides and adenosine formation in the isolated urothelium from organ donors (control individuals). ATP and AMP (30 µM) were added at zero time either to the luminal or to the serosal side of the human urothelium. Samples (75 µL) were collected from the incubation fluid at the indicated times on the abscissa. Each collected sample was analysed by HPLC with UV detection to separate and quantify ATP (black), ADP (red), AMP (orange), ADO (green), INO (dark grey) and HX (light grey). Average results obtained in three to five individuals. The calculated half-life time ($t_{½}$) for each initial substrate is shown for comparison.

Interestingly, the basal layer of the urothelium stained most significantly against the ecto-5'-nucleotidase antibody. Tissue elements below the urothelium (e.g. blood vessels, myofibroblasts and nerve fibers), as well as the detrusor smooth muscle layer, were positively stained with NTPDase1, NTPDase2 and ecto-5'-nucleotidase suggesting that biosynthesis of adenosine from adenine nucleotides is positioned to favor a more important role of the nucleoside in the suburothelium and detrusor smooth muscle (Figure 1).

The relative contribution of distinct ectonucleotidases to the modulation of purinergic signalling depends on differential tissue distribution, regulation of cell expression, targeting to specific membrane domains, but also on substrate preference and availability. This might explain the different kinetics of ATP metabolism and adenosine accumulation between luminal and serosal sides of the urothelium (Lewis and Lewis, 2006). Ussing chamber experiments with the rabbit urothelium showed that adenosine concentration rises in the incubation fluid by stretching the tissue, providing that the experiments are done in the presence of iodotubercidin (an inhibitor of AK) and erythro-9-(2-hydroxy-3-nonyl)adenine (EHNA, an inhibitor of ADA) (Yu et al., 2006; Prakasam et al., 2012); the highest concentration of adenosine was recorded at the serosal surface of the urothelium, with concentrations approaching 200 nM. Previous studies from the same group showed that increased pressure stimulates ATP release from both surfaces of the uroepithelium, but mucosal release of the nucleotide was 50-fold greater than serosal release (Wang et al., 2005). The disparity between the site of preferential ATP and adenosine accumulation at the two urothelial surfaces led those authors to hypothesized that hydrolysis of released ATP by ectonucleotidases would play a minor role on adenosine biosynthesis in the rabbit urothelium. Yet, this might not be the case if one considers our data showing that the rate limiting enzyme for the extracellular formation of adenosine from released adenine nucleotides, ecto-5'-nucleotidase, is located predominantly at the serosal surface of the urothelium (i.e. basal urothelial cells, suburothelial connective elements and detrusor smooth muscle fibers). Thus, at this stage, one cannot ignore that adenosine formation from ATP released from the urothelium is preferential at the serosal epithelial site (see Figure 1), which together with adenosine released as such via equilibrative nucleoside transporters may contribute to the high levels of adenosine observed in suburothelial layers. The relative contribution of adenosine formation from cyclic AMP produced within the urothelium in response to increased hydrostatic pressure (Wang et al., 2003) via ecto-phosphodiesterases and ecto-5'-nucleotidase requires investigation in further studies.

Fine-tuning regulation of extracellular purines in the urothelium is more complex than first hypothesized. Besides adenosine being originated from the breakdown of released adenine nucleotides via the ectonucleotidase pathway (see above), it has been demonstrated that it may negatively modulate stretched-induced ATP release over many minutes by modulating exocytosis in the apical membrane of umbrella cells (Yu et al., 2006). A more acute (within seconds) effect to modulate distension-induced ATP release from the urothelium has been suggested (Dunning-Davies et al., 2013). Distension-induced ATP release was decreased by adenosine (1-10 μM) and enhanced by ADA and by the A_1 receptor antagonist, DPCPX, but not by blocking A_2 receptors with 3,7-dimethyl-1-propargyl-xanthine (DMPX). The mechanism by which adenosine reduces urothelial ATP release remains unknown, but it might depend on changes in the transepithelial potential by interference with mechanosensitive Na^+ channels (identified as amiloride-sensitive epithelial Na^+ channels or ENaC), which may ultimately favor Ca^{2+} outflow via Na^+/Ca^{2+} exchange in urothelial cells (Wu et al., 2011). These findings support a dominant role of A_1 receptors in regulating ATP release from the urothelium. This information, which was obtained in rabbit umbrella cells, is in harmony with rat cystometry data showing that luminal application of the A_1 receptor agonist, CCPA, significantly decreased the frequency of the micturition reflex in clear opposition to that observed upon increasing ATP concentration inside the bladder (Kitta et al., 2013). Conflicting results, however, exist as adenosine acting probably via A_1 receptors decreased the threshold pressure needed to stimulate the micturition reflex, thus increasing bladder hyperactivity, in a model of cyclophosphamide-induced cystitis (Prakasam et al., 2012). Whether adenosine exerts diverse effects to control bladder function under pathological conditions versus control situations through coupling to distinctive intracellular signaling pathways, needs clarification.

Whereas adenosine binds to its receptors and impacts organ-protective functions (Cohen and Downey, 2008), chronic elevation of extracellular adenosine is harmful to tissues (Zhou et al., 2010). Hence, the half-life time of extracellular adenosine is kept short (seconds to minutes, depending on species) through the action of enzymes and transporters. Taking into consideration that adenosine levels at the luminal site of the urothelium are 10-fold below those found in the serosal side, one may hypothesize that the adenosine inactivation mechanisms are particularly active at the luminal surface of the bladder; this has also relevance to keep low urinary adenosine originated upstream the bladder urinary tract (e.g. kidney) (Jackson et al., 2011). In this respect, ADA seems to be most important enzyme to control adenosine levels at the mucosal surface of the bladder, while adenosine kinase and nucleoside transporters seem to play more significant roles at the serosal side of the uroepithelial cell layer (Prakasam et al., 2012). This hypothesis gains further support considering previous studies showing that the basolateral surface of polarized epithelial cells are endowed with equilibrative nucleoside transporters (Loffler et al., 2007).

Effects of Adenosine at the Detrusor Neuromuscular Junction

The adenosine content of human bladder smooth muscle is 6.7 times higher than striated muscle and the adenosine/ATP ratio is 1:9 compared with 1:450 for striated muscle (Wedenberg et al., 1994). Several studies have shown that adenosine concentration-dependently relaxes pre-contracted urinary bladder detrusor strips exposed to carbachol, acetylcholine (ACh), potassium depolarization, or electrical field stimulation in rats, guinea-pigs (Acevedo et al., 1992; Brown et al., 1979; Burnstock et al., 1978a; King et al., 1997; Nicholls et al., 1992) and humans (Rubinstein et al., 1998). Maximum relaxation by adenosine never exceeded 50% of ACh-induced tension in the isolated human detrusor (Rubinstein et al., 1998). Despite increases in the bladder purinergic tone (ATP-induced contractions) have been associated with interstitial cystitis, detrusor overactivity, outlet obstruction, inflammation, neurogenic bladder, spinal cord lesions and aging, we were unable to detected any measurable increase in the sensitivity to the relaxant effect of adenosine in hypertrophied bladders secondary to partial outlet obstruction due to prostatic adenoma as compared to the control group (organ donors) (unpublished observations).

Characterization of the adenosine receptor subtype responsible for detrusor relaxation has been difficult to establish due to differences in target species and experimental protocols. Based on the rank potency order of P1 agonists to reduce contractile responses caused by ATP or ACh, it seems that an A_2 receptor subtype might be involved (Acevedo et al., 1992; Nicholls et al., 1992). Characterization of the A_2 receptor subtype implicated in this process is hindered by the absence of highly selective and potent ligands for A_{2B} receptors. In the guinea-pig bladder smooth muscle cells adenosine evokes membrane hyperpolarization by activation of K_{ATP} channels involving activation of adenylate cyclase, generation of cyclic AMP and stimulation of protein kinase A (Gopalakrishnan et al., 2002). In this study, the non-selective A_2 receptor agonist, 5'-*N*-ethylcarboxamidoadenosine (NECA), was equipotent to the selective A_{2A} receptor agonist, CGS21680C, and these were more potent than A_1 and A_3 selective agonists, 2-choloro-N^6-cylcopentyladenosine (CCPA) and 2-chloro-N^6-(3-iodobenzyl) adenosine-5'-N-methylcarboxamide (2-Cl-IBMECA), respectively. Moreover, the selective A_{2A} receptor antagonist, ZM241385, inhibited the changes in membrane potential evoked by adenosine and NECA on detrusor smooth muscle cells. However, There there is are also however conflicting data with the use of the A_{2A} receptor agonist CGS21680C as it relaxed the guinea-pig detrusor strips pre-contracted with KCl (Gopalakrishnan et al., 2002) but caused only a slight inhibition compared to NECA on carbachol-induced contractions in the urinary bladder of rats (Nicholls et al., 1992) and pigs (Hernandez et al., 1999). This also insinuates a role for low affinity A_{2B} receptors in this process. In fact, the A_{2B} selective antagonist, PSB 1115, blocked the relaxant effect of adenosine when it was applied in high concentrations (Aronsson et al., 2010). Thus, so far there is no clear cut evidence to decide which A_2 receptor subtype predominantly participates in adenosine relaxation of the detrusor muscle (question mark in Figure 2).

Other authors (Vesela et al., 2011) claimed that the A_1 receptor may also partially operate adenosine-induced relaxation of the rat urinary bladder (acting at the detrusor and urothelium), since the selective A_1 receptor antagonist, 1,3-dipropyl-8-cyclopentylxanthine (DPCPX), but not the A_{2A} receptor antagonist, SCH58261, blocked the effect of the nucleoside. Conversely, findings obtained under different experimental conditions showed that the A_1 receptor may be responsible for contraction of isolated detrusor cells from cats by acting via pertussis toxin-sensitive G_{i3} protein, phospholipase C-β_3 and the release of intracellular Ca^{2+} (Yang et al., 2000). Controversy regarding the effects of adenosine in the urinary bladder detrusor increases given that data demonstrate that blockage of A_3 receptors may potentiate adenosine relaxation, suggesting that stimulation of A_3 receptors induces detrusor contraction in the rat (Vesela et al., 2011).

Besides direct actions of adenosine on P1 receptors located on detrusor smooth muscle fibers, the nucleoside and its analogues may act indirectly by inhibiting nerve-evoked contractions; this was evidenced in rats (Acevedo et al., 1992; Burnstock et al., 1978a; King et al., 1997; Parija et al., 1991), guinea-pigs (Burnstock et al., 1978a), cats (Yang et al., 2000) and humans (Husted et al., 1983). Interestingly, inhibition of evoked transmitters release by adenosine is much more pronounced than the effect of the nucleoside on pre-contracted human detrusor smooth muscle (Figure 2).

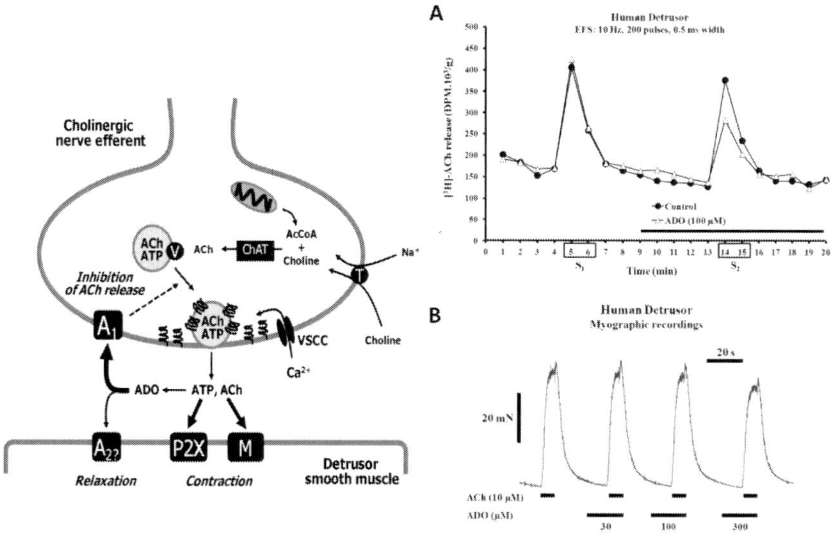

Figure 2. Inhibition of neuromuscular transmission by adenosine in the human detrusor is due to a preferential activation of A_1 receptors on cholinergic nerve efferents, while the nucleoside produces less potent relaxation by directly targeting receptors on smooth muscle fibers. Panel A, shows the time course of tritium outflow from isolated human detrusor strips taken from typical experiments in the absence (Control, filled circles) and in the presence of adenosine (ADO, open triangles). The abscissa indicates the times at which samples were collected. [^{3}H]-ACh release was elicited twice (S_1 and S_2) by electrical field stimulation (10 Hz, 200 pulses of 0.510.5 ms duration) of the preparations. ADO (100 μM) was applied 15 min before the end of S_2, as represented by the horizontal bar. Panel B, shows typical myographic recording traces obtained during contractile responses of the isolated human detrusor exposed to ACh (10 μM) in the absence and in the presence of ADO. ADO (30, 100 and 300 μM) was applied for 20 min, as represented by the horizontal bar. The small horizontal lines indicate the duration of ACh (10 μM) application; vertical calibration: 20 mN. Note that the concentration of ADO required to reduce ACh (10 μM)-induced detrusor contractions was higher (300 μM) than that needed to decrease evoked [^{3}H]-ACh release; at 100 μM, ADO inhibited by 39±6% (*n*=4) stimulation-evoked transmitter release, yet no relaxing effect was observed on isolated detrusor strips exposed to ACh (10 μM).

The rank potency order of adenosine analogues to reduce nerve-evoked contractions is clearly different from that required to reduce ATP- and/or carbachol-induced contractions. Being A_1 receptor agonists, like N^6-(R-2-phenylisopropyl)adenosine (R-PIA) and N^6-cyclohexyladenosine (CHA), the most potent ligands on electrically-evoked detrusor contractions, whereas NECA was the most potent agonist on detrusor contractions caused by ATP or carbachol (Acevedo et al., 1992; Nicholls et al., 1992). These findings are consistent with idea that adenosine inhibition of nerve-evoked detrusor contractions is related to the activation of pre-synaptic A_1 receptors (Williams, 1987). This is further supported with experiments in isolated human detrusor strips designed to measure [^{3}H]ACh release triggered by electrical field stimulation. Under these conditions, we showed that adenosine and the selective A_1 receptor agonist, R-PIA, significantly reduced [^{3}H]ACh release and that this effect was fully blocked by the A_1 antagonist, DPCPX (Silva-Ramos et al., 2010). Figure 2 shows that adenosine (100 μM) was more potent in reducing nerve-evoked transmitter release than in producing relaxation of muscle strips pre-contracted with ACh (10 μM) in the isolated human detrusor.

Adenosine also seems to have an important role in non-adrenergic non-cholinergic (NANC), probably nitrergic, inhibition of the bladder neck. ATP-induced relaxation of this tissue seems to be mediated by metabotropic $P2Y_1$ and A_{2A} receptors after its breakdown to ADP and adenosine, respectively (Hernandez et al., 2009).

Effects of Adenosine in the Control of Micturition Reflex

Adenosine receptors are widely distributed in the central nervous system being A_1 and A_{2A} receptors the most expressed subtypes (Dixon et al., 1996). The A_1 receptor is widely distributed in the brain, spinal cord and peripheral nervous system (Fredholm et al., 2001), whereas the A_{2A} receptor is predominantly expressed in basal ganglia and in presynaptic terminals of sensory afferents (Dunwiddie and Masino, 2001; Kaelin-Lang et al., 1998).

Most of the evidence for the modulatory role of adenosine on the micturition reflex, both in the periphery and in the central nervous system, comes from cystometry experiments performed in the rat. It has been reported that both R-PIA and NECA adenosine analogues administered intrathecaly delayed the voiding reflex (Sosnowski and Yaksh, 1990). More recently, it was found that activation of A_1 receptors or inhibition A_{2A} receptors in the brain or in the spinal cord cause an inhibitory effect on the micturition reflex (Kitta et al., 2013). These authors demonstrated that amplification of the inhibitory A_1-receptor-mediated effect was observed in the brain of rats with bladder overactivity caused by intravesical instillation of acetic acid, whereas the A_{2A} receptor tonus increased in the spinal cord. A peripheral inhibitory action was observed in this model when A_1 agonists were applied into the lumen of the urinary bladder. Interestingly, A_1 and A_{2A} receptor mediated effects control the micturition reflex acting predominantly on the afferent arm of the reflex, without significant changes in the voiding pressure.

Role of Adenosine on Lower Urinary Tract Symptoms

The physiological effect of adenosine is now considered a new direction in halting the progression of damage in several organ systems. Two of the most common conditions that cause lower urinary tract symptoms (LUTS) are detrusor overactivity and bladder outlet obstruction due to benign prostatic hyperplasia. During the last decades a huge amount of evidences implicate the purinergic system in the pathogenesis of urinary bladder syndromes (Burnstock, 2014).

Overactive bladder (OAB) is defined by the International Continence Society as the presence of urinary urgency, usually accompanied by frequency and nocturia, with or without urinary incontinence, in the absence of a causative infection or obvious pathological conditions (Abrams et al., 2002). Although there is a lot of evidence supporting the role of ATP in the pathogenesis of OAB, surprisingly there is few data on the role of its metabolite, adenosine, or its absence in the pathogenesis of this syndrome. Nevertheless, we found a study in a rat model of Parkinson disease suggesting that bladder overactivity in this condition was, at least in part, caused by increased activity of A_{2A} receptors in the brain (Kitta et al., 2012). This is not surprising given that these receptors are selectively co-localized and negatively interplay with dopamine D_2 receptors on GABAergic output neurons of the striatopallidal pathway (Mori and Shindou, 2003). Moreover, the inhibitory effects of adenosine on ATP release from urothelial cells, on cholinergic nerve activity in the detrusor and, directly, on detrusor smooth muscle contraction, envisage some role of the nucleoside in the pathophysiology of OAB. In clinical practice, the mainstay for treatment of OAB is anticholinergic drugs, such as oxybutinin. However, more than half of patients drop out medication within the first 3 months because of lack of effect, adverse events and/or cost of treatment (Gopal et al., 2008). A recent study has shown that chronic administration of oxybutynin induces a shift from muscarinic to purinergic transmission in the bladder wall of rats and it was suggested that this may partially explain the high discontinuation rate of anticholinergics used for the treatment of OAB (Uvin et al., 2013). More recently, mirabegron, a new β_3 adrenoceptor agonist (Igawa and Michel, 2013) has been developed as a bladder relaxant used in the treatment of OAB syndrome. Currently, no drug acting on purinoceptors is available for the treatment of diseases in the lower urinary tract; even though work has been carry out to identify substances acting on purinoceptors located in sensory neurons of the human urinary bladder (Gever et al., 2010). In the quest for new treatment options for OAB, modulation of endogenous adenosine levels and/or P1 receptors activity stands as an interesting prospect.

Bladder outlet obstruction due to benign prostatic hyperplasia (BPH) induces several alterations on the detrusor morphology and function (Mirone et al., 2007). Detrusor from these patients usually exhibit a greater component of atropine resistance and an enhanced release of ATP and ACh when electrically stimulated (Bayliss et al., 1999; Silva-Ramos et al., 2009). Adenosine has a much more pronounced pre-synaptic inhibitory effect via A_1 receptors decreasing ACh release from cholinergic nerve efferents innervating the detrusor of these patients as compared to controls (Silva-Ramos et al., 2010). This was verified either by using selective adenosine analogues, like R-PIA, or by showing that adenosine inhibition of transmitter release was significantly impaired by blockage of A_1 receptors with DPCPX or caffeine. Most importantly, we demonstrated that the adenosine A_1 receptor-mediated inhibitory tonus may be strengthen by inhibiting endogenous adenosine inactivation, by cellular uptake (with dipyridamole) or deamination (with EHNA), which can be useful in the treatment of patients with bladder overactivity given that no direct changes were observed in detrusor contractile tension that would otherwise undermine voiding in these patients (see above).

Painful bladder syndrome/interstitial cystitis (BPS/IC) although not so frequent as OAB and BPH, has a huge impact on quality of life and till now, no effective treatment (Theoharides et al., 2008). BPS/IC is a chronic inflammation of bladder subendothelium and muscular tissues of the urinary bladder not caused by an infectious organism (van de Merwe et al., 2008). Regarding purinergic alterations in BPS/IC, it has been demonstrated that urinary ATP levels are elevated and that the release of ATP to stretch of urothelial cells is increased in patients with this disease (Sun and Chai, 2006). *In vitro* examinations of bladder strip preparations from patients with and without BPS/IC have suggested that the purinergic signaling may be minute, or even absent, in healthy humans (Kinder and Mundy, 1985), but of markedly increased importance in patients with this disorder (Palea et al., 1993). Several studies suggest the involvement of P1 receptors in the regulation of the inflammatory response (Hasko et al., 2008; Sitkovsky et al., 2004). In rats with cyclophosphamide induced cystitis, the A_1 receptor activation mediates important pro-inflammatory responses (Aronsson et al., 2012) and paradoxical voiding effects (Prakasam et al., 2012; see above). In fact, blockade of A_1 receptors with DPCPX normalized bladder contractile responses and prevented inflammatory bladder wall changes induced by cyclophosphamide. The A_1 receptor-mediated relaxation is perhaps replaced by A_{2B} receptors function, while A_3 receptors mediate contraction (Vesela et al., 2011). On the other hand, adenosine activation of A_{2A} has been shown to provide potent anti-inflammatory effects and A_{2A} receptor are highly expressed in immune cells such as monocyte/macrophages, neutrophils and T cells (see e.g. Sitkovsky et al., 2004). In human urothelial cells infected with *Escherichia coli,* the expression of A_1 and A_{2B} mRNA was decreased and the expression of A_{2A} was increased suggesting that adenosine may regulate inflammatory responses in response to infection via A_{2A} receptors (Save et al., 2009).

Conclusion

Although innumerous studies demonstrate the pathophysiological relevance of purines, namely ATP and related nucleotides, in the urinary bladder, controversy still exists about the main signaling role of their major extracellular metabolite, adenosine. More studies are required to fill the gap in our knowledge regarding the pathophysiological implications of the nucleoside in normal and pathological human bladder function. Data from several groups mostly using animal models demonstrated that adenosine regulates micturition by controlling the mechanosensitive interplay between the urothelium and low threshold suburothelial nerve afferents, through the release of ATP. The preferential biosynthesis of adenosine in suburothelial and detrusor smooth muscle layers, affords for a predominant inhibitory tone of the nucleoside controlling sensory nerves firing and ACh release from cholinergic nerve efferents. Most of the studies point towards the predominant involvement of A_1 receptors in the inhibitory action of adenosine at the urinary bladder. At the human detrusor neuromuscular junction, adenosine inhibition is operated by A_1 receptors localized on cholinergic nerve terminals, without much affecting the contractility of the detrusor when the nucleoside was used in physiological concentrations.

Adenosine also impacts the control of the micturition reflex mediated by central nervous centers via the activation of both inhibitory A_1 and excitatory A_{2A} receptors, as it has been evidenced in animal models of urinary disorders. If the role of the purinoceptors in direct bladder function is, or has been, considered to be a bit unorthodox, the fact that ATP and adenosine, by acting on their respective purinoceptors, are able to affect the course of inflammation is less controversial. It has been demonstrated that the anti-inflammatory and immunosuppressive action of A_{2A} receptors on infiltrating immune cells may be relevant in BPS/IC.

Data from our group demonstrate that bladder overactivity symptoms subsequent to bladder outlet obstruction due to prostatic adenoma may be owe to increased accumulation of ATP and consequent decreased adenosine formation due to reduced extracellular hydrolysis of the nucleotide by NTPDases. While this grants increased access of nucleotides to excitatory P2 purinoceptors located on sensory nerve afferents (P2X2/3 or P2X3) and smooth muscle fibers (P2X1), low endogenous levels of adenosine releases the break that controls bladder excitation (at urothelial cells, suburotelial nerves and cholinergic nerve efferents) yielding leading to a compensatory upregulation of inhibitory A_1 receptors (unpublished observations). Within this context, it is tempting to hypothesize that drugs targeting the inhibitory A_1 receptor or that bolster endogenous adenosine accumulation (e.g. nucleoside uptake blockers and/or ADA inhibitors) might have potential benefits in the treatment of lower urinary tract syndromes, in parallel to the identified P2X receptor antagonists acting on sensory neurons of the bladder (Gever et al., 2010). Based on the results obtained in animal studies *in vivo*, administration of these drugs by instillation into the bladder may be an option to systemic administration in order to circumvent putative adverse effects, using a strategy inspired by that applied for antimuscarinic drugs.

Acknowledgments

This research was partially supported by Fundação para a Ciência e a Tecnologia (FCT projects PTDC/SAU-OSM/104369/2008 and Pest/OE/UI0215/2014), Associação Portuguesa de Urologia and University of Porto / Caixa Geral de Depósitos (Investigação Científica na Pré-Graduação).

Statement of Conflicts of Interest: None.

References

Abrams P, Cardozo L, Fall M, Griffiths D, Rosier P, Ulmsten U, van Kerrebroeck P, Victor A, Wein A. 2002. The standardisation of terminology of lower urinary tract function: report from the Standardisation Sub-committee of the International Continence Society. *Neurourol. Urodyn* 21(2):167-178.

Acevedo CG, Contreras E, Escalona J, Lewin J, Huidobro-Toro JP. 1992. Pharmacological characterization of adenosine A1 and A2 receptors in the bladder: evidence for a modulatory adenosine tone regulating non-adrenergic non-cholinergic neurotransmission. *Br. J. Pharmacol.*. 107(1):120-126.

Apodaca G. 2004. The uroepithelium: not just a passive barrier. *Traffic* 5(3):117-128.

Apodaca G, Balestreire E, Birder LA. 2007. The uroepithelial-associated sensory web. *Kidney Int.* 72(9):1057-1064.

Arch JR, Newsholme EA. 1978. The control of the metabolism and the hormonal role of adenosine. *Essays Biochem.* 14:82-123.

Aronsson P, Andersson M, Ericsson T, Giglio D. 2010. Assessment and characterization of purinergic contractions and relaxations in the rat urinary bladder. *Basic Clin. Pharmacol. Toxicol.* 107(1):603-613.

Aronsson P, Johnsson M, Vesela R, Winder M, Tobin G. 2012. Adenosine receptor antagonism suppresses functional and histological inflammatory changes in the rat urinary bladder. *Auton Neurosci.* 171(1-2):49-57.

Bayliss M, Wu C, Newgreen D, Mundy AR, Fry CH. 1999. A quantitative study of atropine-resistant contractile responses in human detrusor smooth muscle, from stable, unstable and obstructed bladders. *J. Urol.* 162(5):1833-1839.

Birder L, Andersson KE. 2013. Urothelial signaling. *Physiol. Rev.* 93(2):653-680.

Birder L, de Groat W, Mills I, Morrison J, Thor K, Drake M. 2010. Neural control of the lower urinary tract: peripheral and spinal mechanisms. *Neurourol. Urodyn.* 29(1):128-139.

Birder LA. 2005. More than just a barrier: urothelium as a drug target for urinary bladder pain. *Am. J. Physiol. Renal Physiol.* 289(3):F489-495.

Birder LA, de Groat WC. 2007. Mechanisms of disease: involvement of the urothelium in bladder dysfunction. *Nat. Clin. Pract. Urol.* 4(1):46-54.

Birder LA, Ruggieri M, Takeda M, van Koeveringe G, Veltkamp S, Korstanje C, Parsons B, Fry CH. 2012. How does the urothelium affect bladder function in health and disease? ICI-RS 2011. *Neurourol. Urodyn.* 31(3):293-299.

Brown C, Burnstock G, Cocks T. 1979. Effects of adenosine 5'-triphosphate (ATP) and beta-gamma-methylene ATP on the rat urinary bladder. *Br. J. Pharmacol.* 65(1):97-102.

Burnstock G. 2014. Purinergic signalling in the urinary tract in health and disease. *Purinergic Signal.* 10(1):103-155.

Burnstock G, Cocks T, Crowe R, Kasakov L. 1978a. Purinergic innervation of the guinea-pig urinary bladder. *Br. J. Pharmacol.*. 63(1):125-138.

Burnstock G, Cocks T, Kasakov L, Wong HK. 1978b. Direct evidence for ATP release from non-adrenergic, non-cholinergic ("purinergic") nerves in the guinea-pig taenia coli and bladder. *Eur. J. Pharmacol.*. 49(2):145-149.

Burnstock G, Dumsday B, Smythe A. 1972. Atropine resistant excitation of the urinary bladder: the possibility of transmission via nerves releasing a purine nucleotide. *Br. J. Pharmacol.*. 44(3):451-461.

Cohen MV, Downey JM. 2008. Adenosine: trigger and mediator of cardioprotection. *Basic Res. Cardiol.* 103(3):203-215.

Colgan SP, Eltzschig HK, Eckle T, Thompson LF. 2006. Physiological roles for ecto-5'-nucleotidase (CD73). *Purinergic. Signal.* 2(2):351-360.

Correia-de-Sá P. 2010. Pathophysiological significance of nucleotide hydrolyzing enzymes in urologic disorder. *Purinergic Signal.* 6(Suppl 2):S13.13.

Correia-de-Sá P, Adaes Adães S, Timoteo Timóteo MA, Vieira C, MagalhaesMagalhães-Cardoso T, Nascimento C, Duarte-Araujo Araújo M. 2006. Fine-tuning modulation of myenteric motoneurons by endogenous adenosine: on the role of secreted adenosine deaminase. *Auton. Neurosci.* 126-127:211-224.

Correia-de-Sá P, Ribeiro JA. 1996. Adenosine uptake and deamination regulate tonic A_{2a} receptor facilitation of evoked [^{3}H]acetylcholine release from the rat motor nerve terminals. *Neuroscience* 73(1):85-92.

Cusack NJ, Hourani SM. 1984. Some pharmacological and biochemical interactions of the enantiomers of adenylyl 5'-(beta, gamma-methylene)-diphosphonate with the guinea-pig urinary bladder. *Br. J. Pharmacol.*. 82(1):155-159.

Dixon AK, Gubitz AK, Sirinathsinghji DJ, Richardson PJ, Freeman TC. 1996. Tissue distribution of adenosine receptor mRNAs in the rat. *Br. J. Pharmacol.* 118(6):1461-1468.

Duarte-Araujo Araújo M, Nascimento C, Alexandrina Timoóteo MA, MagalhaesMagalhães-Cardoso T, Correia-de-Sá P. 2004. Dual effects of adenosine on acetylcholine release from myenteric motoneurons are mediated by junctional facilitatory A(2A) and extrajunctional inhibitory A(1) receptors. *Br. J. Pharmacol.* 141(6):925-934.

Dunning-Davies BM, Fry CH, Mansour D, Ferguson DR. 2013. The regulation of ATP release from the urothelium by adenosine and transepithelial potential. *BJU Int.* 111(3):505-513.

Dunwiddie TV, Masino SA. 2001. The role and regulation of adenosine in the central nervous system. *Annu. Rev. Neurosci.* 24:31-55.

Ferguson DR, Kennedy I, Burton TJ. 1997. ATP is released from rabbit urinary bladder epithelial cells by hydrostatic pressure changes--a possible sensory mechanism? *J. Physiol.* 505 (Pt 2):503-511.

Fredholm BB, AP IJ, Jacobson KA, Klotz KN, Linden J. 2001. International Union of Pharmacology. XXV. Nomenclature and classification of adenosine receptors. *Pharmacol. R*ev. 53(4):527-552.

Geiger JF, Fyda DM. 1991. Adenosine in the Nervous System: Adenosine transport in nervous system tissues. *United States Edition*. San Diego: Academic Press Inc.

Gendron FP, Benrezzak O, Krugh BW, Kong Q, Weisman GA, Beaudoin AR. 2002. Purine signaling and potential new therapeutic approach: possible outcomes of NTPDase inhibition. *Curr. Drug Targets* 3(3):229-245.

Gever JR, Soto R, Henningsen RA, Martin RS, Hackos DH, Panicker S, Rubas W, Oglesby IB, Dillon MP, Milla ME, Burnstock G, Ford AP. 2010. AF-353, a novel, potent and orally bioavailable P2X3/P2X2/3 receptor antagonist. *Br. J. Pharmacol.* 160(6):1387-1398.

Gillespie JI, Markerink-van Ittersum M, de Vente J. 2006. Sensory collaterals, intramural ganglia and motor nerves in the guinea-pig bladder: evidence for intramural neural circuits. *Cell Tissue Res.* 325(1):33-45.

Goncalves J, Queiroz G. 1993. Facilitatory and inhibitory modulation by endogenous adenosine of noradrenaline release in the epididymal portion of rat vas deferens. *Naunyn Schmiedeberg's Arch. Pharmacol.* 348(4):367-371.

Gopal M, Haynes K, Bellamy SL, Arya LA. 2008. Discontinuation rates of anticholinergic medications used for the treatment of lower urinary tract symptoms. *Obstet Gynecol.* 112(6):1311-1318.

Gopalakrishnan SM, Buckner SA, Milicic I, Groebe DR, Whiteaker KL, Burns DJ, Warrior U, Gopalakrishnan M. 2002. Functional characterization of adenosine receptors and coupling to ATP-sensitive K+ channels in Guinea pig urinary bladder smooth muscle. *J. Pharmacol. Exp. Ther.* 300(3):910-917.

Hasko G, Linden J, Cronstein B, Pacher P. 2008. Adenosine receptors: therapeutic aspects for inflammatory and immune diseases. *Nat. Rev. Drug Discov.* 7(9):759-770.

Hernandez M, Barahona MV, Bustamante S, Garcia-Sacristan A, Orensanz LM. 1999. A_{2B} adenosine receptors mediate relaxation of the pig intravesical ureter: adenosine modulation of non adrenergic non cholinergic excitatory neurotransmission. *Br. J. Pharmacol.* 126(4):969-978.

Hernandez M, Knight GE, Wildman SS, Burnstock G. 2009. Role of ATP and related purines in inhibitory neurotransmission to the pig urinary bladder neck. *Br. J. Pharmacol.* 157(8):1463-1473.

Heyne N, Benohr P, Muhlbauer B, Delabar U, Risler T, Osswald H. 2004. Regulation of renal adenosine excretion in humans--role of sodium and fluid homeostasis. *Nephrol. Dial Transplant* 19(11):2737-2741.

Husted S, Sjogren C, Andersson KE. 1983. Direct effects of adenosine and adenine nucleotides on isolated human urinary bladder and their influence on electrically induced contractions. *J. Urol.* 130(2):392-398.

Igawa Y, Michel MC. 2013. Pharmacological profile of beta3-adrenoceptor agonists in clinical development for the treatment of overactive bladder syndrome. *Naunyn Schmiedebergs Arch. Pharmacol.* 386(3):177-183.

Irwin DE, Milsom I, Hunskaar S, Reilly K, Kopp Z, Herschorn S, Coyne K, Kelleher C, Hampel C, Artibani W, Abrams P. 2006. Population-based survey of urinary incontinence, overactive bladder, and other lower urinary tract symptoms in five countries: results of the EPIC study. *Eur. Urol.* 50(6):1306-1314; discussion 1314-1305.

Jackson EK, Raghvendra DK. 2004. The extracellular cyclic AMP-adenosine pathway in renal physiology. *Annu. Rev. Physiol.* 66:571-599.

Jackson EK, Ren J, Cheng D, Mi Z. 2011. Extracellular cAMP-adenosine pathways in the mouse kidney. *Am. J. Physiol. Renal Physiol.* 301(3):F565-573.

Kaelin-Lang A, Lauterburg T, Burgunder JM. 1998. Expression of adenosine A_{2a} receptor gene in rat dorsal root and autonomic ganglia. *Neurosci. Lett.* 246(1):21-24.

Kinder RB, Mundy AR. 1985. Atropine blockade of nerve-mediated stimulation of the human detrusor. *Br. J. Urol.* 57(4):418-421.

King JA, Huddart H, Staff WG. 1997. Purinergic modulation of rat urinary bladder detrusor smooth muscle. *Gen. Pharmacol.* 29(4):597-604.

Kitta T, Chancellor MB, de Groat WC, Kuno S, Nonomura K, Yoshimura N. 2012. Suppression of bladder overactivity by adenosine A_{2A} receptor antagonist in a rat model of Parkinson disease. *J. Urol.* 187(5):1890-1897.

Kitta T, Chancellor MB, de Groat WC, Kuno S, Nonomura K, Yoshimura N. 2013. Roles of adenosine A_1 and A_{2A} receptors in the control of micturition in rats. *Neurourol. Urodyn.* Doi:10.1002/nau.22487.

Klinger M, Freissmuth M, Nanoff C. 2002. Adenosine receptors: G protein-mediated signalling and the role of accessory proteins. *Cell Signal.* 14(2):99-108.

Kukulski F, Levesque SA, Lavoie EG, Lecka J, Bigonnesse F, Knowles AF, Robson SC, Kirley TL, Sevigny J. 2005. Comparative hydrolysis of P2 receptor agonists by NTPDases 1, 2, 3 and 8. *Purinergic Signal.* 1(2):193-204.

Lewis SA, Lewis JR. 2006. Kinetics of urothelial ATP release. *Am. J. Physiol. Renal Physiol.* 291(2):F332-340.

Lloyd HG, Fredholm BB. 1995. Involvement of adenosine deaminase and adenosine kinase in regulating extracellular adenosine concentration in rat hippocampal slices. *Neurochem. Int.* 26(4):387-395.

Loffler M, Morote-Garcia JC, Eltzschig SA, Coe IR, Eltzschig HK. 2007. Physiological roles of vascular nucleoside transporters. *Arterioscler Thromb. Vasc. Biol.* 27(5):1004-1013.

Maggi CA. 1990. The dual function of capsaicin-sensitive sensory nerves in the bladder and urethra. *Ciba Found Symp.* 151:77-83; discussion 83-90.

Mirone V, Imbimbo C, Longo N, Fusco F. 2007. The detrusor muscle: an innocent victim of bladder outlet obstruction. *Eur. Urol.* 51(1):57-66.

Mori A, Shindou T. 2003. Modulation of GABAergic transmission in the striatopallidal system by adenosine A_{2A} receptors: a potential mechanism for the antiparkinsonian effects of A_{2A} antagonists. *Neurology* 61(11 Suppl 6):S44-48.

Nicholls J, Hourani SM, Kitchen I. 1992. Characterization of P1-purinoceptors on rat duodenum and urinary bladder. *Br. J. Pharmacol.* 105(3):639-642.

Owen SJ, Massa HH, Rose'Meyer RB. 2012. Loss of adenosine A_{2B} receptor mediated relaxant responses in the aged female rat bladder; effects of dietary phytoestrogens. *Naunyn Schmiedeberg's Arch. Pharmacol.* 385(5):539-549.

Palea S, Artibani W, Ostardo E, Trist DG, Pietra C. 1993. Evidence for purinergic neurotransmission in human urinary bladder affected by interstitial cystitis. *J. Urol.* 150(6):2007-2012.

Parija SC, Raviprakash V, Mishra SK. 1991. Adenosine- and alpha,beta-methylene ATP-induced differential inhibition of cholinergic and non-cholinergic neurogenic responses in rat urinary bladder. *Br. J. Pharmacol.* 102(2):396-400.

Phelps PT, Anthes JC, Correll CC. 2006. Characterization of adenosine receptors in the human bladder carcinoma T24 cell line. *Eur. J. Pharmacol.* 536(1-2):28-37.

Prakasam HS, Herrington H, Roppolo JR, Jackson EK, Apodaca G. 2012. Modulation of bladder function by luminal adenosine turnover and A1 receptor activation. *Am. J. Physiol. Renal Physiol.* 303(2):F279-292.

Robson SC, Sevigny J, Zimmermann H. 2006. The E-NTPDase family of ectonucleotidases: Structure function relationships and pathophysiological significance. *Purinergic Signal.* 2(2):409-430.

Rocha I, Burnstock G, Spyer KM. 2001. Effect on urinary bladder function and arterial blood pressure of the activation of putative purine receptors in brainstem areas. *Auton Neurosci.* 88(1-2):6-15.

Rubinstein R, Shalev M, Nissenkorn I, Cohen S. 1998. Effect of exogenous adenosine and its monophosphate on the contractile response to acetylcholine in the human isolated detrusor muscle strips. *J. Auton Pharmacol.* 18(2):99-104.

Save S, Mjosberg J, Poljakovic M, Mohlin C, Persson K. 2009. Adenosine receptor expression in Escherichia coli-infected and cytokine-stimulated human urinary tract epithelial cells. *BJU Int.* 104(11):1758-1765.

Schulte G, Fredholm BB. 2003. Signalling from adenosine receptors to mitogen-activated protein kinases. *Cell Signal.* 15(9):813-827.

Silva-Ramos M, Duarte-Araújo M, Silva I, Lafuente-De-Carvalho JM, Correia-Dede-SÁ P. 2009. Activation of P2X×1 purinoceptors facilitate evoked acetylcholine release in obstructed human bladders. *European Eur. Urology Urol. Supplements Suppl.* 8(4).

Silva-Ramos M, Silva I, Duarte-Araújo M, Magalhães- Cardoso T, Lafuente-de-Ccarvalho J, Correia-de-Sá P. 2010. Loss of balance between excitatory P2X3 and inhibitory A_1 receptors contribute to cholinergic hyperactivity in the obstructed human bladder. *Basic Clin. Pharmacol. Toxicol.* 107((Suppl.1)):130.

Silva-Ramos M, Silva I, Oliveira O, Ferreira S, Reis MJ, Oliveira JC, Correia-de-Sá P. 2013. Urinary ATP may be a dynamic biomarker of detrusor overactivity in women with overactive bladder syndrome. *PLoS One* 8(5):e64696.

Sitkovsky MV, Lukashev D, Apasov S, Kojima H, Koshiba M, Caldwell C, Ohta A, Thiel M. 2004. Physiological control of immune response and inflammatory tissue damage by hypoxia-inducible factors and adenosine A_{2A} receptors. *Annu. Rev. Immunol.* 22:657-682.

Smith DO, Lu Z. 1991. Adenosine derived from hydrolysis of presynaptically released ATP inhibits neuromuscular transmission in the rat. *Neurosci. Lett.* 122(2):171-173.

Sosnowski M, Yaksh TL. 1990. The role of spinal and brainstem adenosine receptors in the modulation of the volume-evoked micturition reflex in the unanesthetized rat. *Brain Res.* 515(1-2):207-213.

Stefan C, Jansen S, Bollen M. 2005. NPP-type ectophosphodiesterases: unity in diversity. *Trends Biochem. Sci.* 30(10):542-550.

Stehle JH, Rivkees SA, Lee JJ, Weaver DR, Deeds JD, Reppert SM. 1992. Molecular cloning and expression of the cDNA for a novel A2-adenosine receptor subtype. *Mol. Endocrinol.* 6(3):384-393.

Sui GP, Wu C, Roosen A, Ikeda Y, Kanai AJ, Fry CH. 2008. Modulation of bladder myofibroblast activity: implications for bladder function. *Am. J. Physiol. Renal Physiol.* 295(3):F688-697.

Sun Y, Chai TC. 2006. Augmented extracellular ATP signaling in bladder urothelial cells from patients with interstitial cystitis. *Am. J. Physiol. Cell Physiol.* 290(1):C27-34.

Theoharides TC, Whitmore K, Stanford E, Moldwin R, O'Leary MP. 2008. Interstitial cystitis: bladder pain and beyond. *Expert Opin. Pharmacother.* 9(17):2979-2994.

Timoteo Timóteo MA, Carneiro I, Silva I, Noronha-Matos JB, Ferreirinha F, Silva-Ramos M, Correia-de-Sá P. 2014. ATP released via pannexin-1 hemichannels mediates bladder overactivity triggered by urothelial $P2Y_6$ receptors. *Biochem. Pharmacol.* 87(2):371-379.

Truschel ST, Wang E, Ruiz WG, Leung SM, Rojas R, Lavelle J, Zeidel M, Stoffer D, Apodaca G. 2002. Stretch-regulated exocytosis/endocytosis in bladder umbrella cells. *Mol. Biol. Cell* 13(3):830-846.

Uvin P, Boudes M, Menigoz A, Franken J, Pinto S, Gevaert T, Verplaetse R, Tytgat J, Vennekens R, Voets T, De Ridder D. 2013. Chronic administration of anticholinergics in rats induces a shift from muscarinic to purinergic transmission in the bladder wall. *Eur. Urol.* 64(3):502-510.

van de Merwe JP, Nordling J, Bouchelouche P, Bouchelouche K, Cervigni M, Daha LK, Elneil S, Fall M, Hohlbrugger G, Irwin P, Mortensen S, van Ophoven A, Osborne JL, Peeker R, Richter B, Riedl C, Sairanen J, Tinzl M, Wyndaele JJ. 2008. Diagnostic criteria, classification, and nomenclature for painful bladder syndrome/interstitial cystitis: an ESSIC proposal. *Eur. Urol.* 53(1):60-67.

Vesela R, Aronsson P, Tobin G. 2011. Functional and morphological examinations of P1A1 purinoceptors in the normal and inflamed urinary bladder of the rat. *Auton Neurosci.* 159(1-2):26-31.

Vidotto C, Fousert D, Akkermann M, Griesmacher A, Muller MM. 2003. Purine and pyrimidine metabolites in children's urine. *Clin. Chim. Acta* 335(1-2):27-32.

Wang E, Truschel S, Apodaca G. 2003. Analysis of hydrostatic pressure-induced changes in umbrella cell surface area. *Methods* 30(3):207-217.

Wang EC, Lee JM, Ruiz WG, Balestreire EM, von Bodungen M, Barrick S, Cockayne DA, Birder LA, Apodaca G. 2005. ATP and purinergic receptor-dependent membrane traffic in bladder umbrella cells. *J. Clin. Invest.* 115(9):2412-2422.

Wedenberg K, Ronquist G, Waldenstrom A, Ulmsten U. 1994. Low energy charge and high adenosine content in smooth muscle of human bladder in comparison with striated muscle. *Int. J. Clin. Lab. Res.* 24(4):230-232.

Williams M. 1987. Purine receptors in mammalian tissues: pharmacology and functional significance. *Annu. Rev. Pharmacol. Toxicol.* 27:315-345.

Yang SJ, An JY, Shim JO, Park CH, Huh IH, Sohn UD. 2000. The mechanism of contraction by 2-chloroadenosine in cat detrusor muscle cells. *J. Urol.* 163(2):652-658.

Yu W, Robson SC, Hill WG. 2011. Expression and distribution of ectonucleotidases in mouse urinary bladder. *PLoS One* 6(4):e18704.

Yu W, Zacharia LC, Jackson EK, Apodaca G. 2006. Adenosine receptor expression and function in bladder uroepithelium. *Am. J. Physiol. Cell Physiol.* 291(2):C254-265.

Zhang SX, Kobayashi T, Okada T, Garcia del Saz E, Seguchi H. 1991. Alkaline phosphatase, 5'-nucleotidase and magnesium-dependent adenosine triphosphatase activities in the transitional epithelium of the rat urinary bladder. *Histol. Histopathol.* 6(3):309-315.

Zhang Y, Xia Y. 2012. Adenosine signaling in normal and sickle erythrocytes and beyond. *Microbes Infect. 14*(10):863-873.

Zhou Y, Murthy JN, Zeng D, Belardinelli L, Blackburn MR. 2010. Alterations in adenosine metabolism and signaling in patients with chronic obstructive pulmonary disease and idiopathic pulmonary fibrosis. *PLoS One* 5(2):e9224.

Zimmermann H. 2000. Extracellular metabolism of ATP and other nucleotides. *Naunyn Schmiedeberg's Arch. Pharmacol.* 362(4-5):299-309.

In: Adenosine Receptors
Editor: Kasandra Warrick

ISBN: 978-1-63463-454-0

Chapter V

Involvement of Deoxyadenosine and Adenosine Deaminase in the Methotrexate-Induced Suppression of Inflammatory Bone Destruction*

Peng-Fei Qu[1], Akiko Kukita[2], Yin-Ji Li[1], Kanako Moriyama[1], Lin Lei[1] and Toshio Kukita[1†]

[1]Department of Molecular Cell Biology & Oral Anatomy, Faculty of Dental Science, Kyushu University, Maidashi, Fuokuoka, Japan

[2]Department of Microbiology, Faculty of Medicine, Saga University, Nabeshima, Saga, Japan

Abstract

Methotrexate (MTX) is an anti-tumor medicine classified into the anti-folate drugs. MTX suppresses DNA synthesis through inhibiting

* Funded in part by a Grant-in-Aid for Scientific Research of Japan, Grant Number 21390492, 21592332, 22659330, 21659424, 24659816.

† Correspondence: Toshio Kukita, Ph.D. Phone: 812-92-642-6300, 6307; Fax: 81-92-642-6304; e-mail: kukitat @ dent.kyushu-u.ac.jp.

dihydrofolate reductase (DHF) of *de novo* nucleic acid pathway, which suppresses proliferation of malignant tumors. MTX is also utilized as for anti-rheumatic medicine as the lower doses of MTX are able to suppress effectively the inflammation of the synovial tissues in RA patients. In our previous study, plasma adenosine level was markedly increased during induction of inflammation in rats with adjuvant-induced arthritis (arthritic rats). MTX-induced suppression of osteoclastogenesis was canceled by the addition of adenosine *in vitro* as well as *in vivo*. Deoxyadenosine (dAdo) is known to regulate proliferation of immune cells and partly share metabolic pathways with adenosine. Adenosine deaminase (ADA) is a key enzyme which metabolize both of adenosine and dAdo to produce inosine and deoxyinosine perpetually. Here we examined the regulatory role of dAdo in MTX-induced suppression of inflammatory bone destruction and investigated a possible involvement of ADA as a target molecule of MTX. In rats with adjuvant-induced arthritis, in whole bone marrow cultures for evaluating osteoclastogenesis, MTX markedly suppressed osteoclastogenesis and dAdo completely canceled its suppression. This cancellation was partially blocked by caffeine, antagonist for the adenosine receptors, A_1AR, A_{2a}AR and A_3AR. Semi-quantitative RT-PCR showed that MTX suppressed expression of RANKL without affecting osteoprotegerin (OPG) expression. Addition of dAdo clearly recovered the expression of RANKL and slightly suppressed OPG expression, which result in augment in osteoclastogenesis. In arthritic rats, MTX-induced suppression of inflammatory bone destruction was completely cancelled by the injection of dAdo. A marked induction of ADA was apparent in bone marrow cells observed in the bone destruction sites around ankle joints in arthritic rats. MTX strikingly suppressed expression of ADA in the corresponding area. Immunoblot analysis also showed a clear induction of ADA protein in arthritic rats and a marked suppression of ADA protein when arthritic rats were treated with MTX. In bone marrow cultures for evaluating osteoclastogenesis, expression of ADA mRNA was markedly suppressed by MTX and dAdo partially cancelled its inhibition. These observations suggest that ADA is involved in the MTX-mediated inhibition of inflammatory bone destruction regulated by adenosine and dAdo. Suppression of ADA by MTX may contribute to the stability of dAdo.

A list of abbreviation: AA, adjuvant-induced arthritis; ADA, adenosine deaminase; AR, adenosine receptor; CFA, complete Freund adjuvant; dAdo, deoxyadenosine; MNC, multinucleated cells; MTX, methotrexate; RA, rheumatoid arthritis; RANK, receptor activator NF-kB ligand, TRAP, tartrate acid-resistant phosphatase.

Keywords: ADA, Adjuvant-induced arthritis, adenosine receptor, Deoxyadenosine, MTX, Osteoclastogenesis

Rheumatoid arthritis (RA) is an inflammatory disease with chronic inflammation in joints which often associates with severe bone destruction [1, 2]. Inflammatory bone destruction is generally accompanied with abnormal increase in osteoclastogenesis in rheumatoid arthritis as well as in periodontal diseases [3].

Osteoclastogenesis is basically controlled by the differentiation factor, the receptor activator NF-κB ligand (RANKL) and the inhibitory factor, osteoprotegerin (OPG) [4]. Inflammatory cytokines are also involved in the pathological augments in osteoclastogenesis [1, 2].

Methotrexate (MTX), a monomethylated form of Aminopterin, is widely utilized as an anti-inflammatory medicine. MTX and Aminopterin have been utilized as the effective anti-tumor drugs as these drugs suppress DNA synthesis through inhibiting the dihydrofolate reductase, a key enzyme required for *de novo* synthesis of nucleic acids [5]. In the area of bone-related diseases, low dose of MTX is utilized to suppress articular inflammation in RA patients [6-9]. However, it has also been recognized that MTX is ineffective to some RA patients [10, 11]. It is likely that MTX suppresses abnormal proliferation of cells present in the synovial membrane observed in RA patients, however, precise target molecule of MTX is ambiguous. Therefore, so far, no one knows the detailed molecular entity of the resistance against MTX.

Intracellular adenosine is the precursor molecule for ATP. Extracellular adenosine is recognized as an important regulator of inflammation [12]. Generally, adenosine is formed from the extracelluar ATP, which is mainly released from apoptotic cells, by a sequential action of cell surface phosphatases, CD39 and CD73 [13].

Extracellular adenosine acts as an inflammatory mediator through cell surface receptors for adenosine. So far four types of adenosine receptors (ARs), A_1AR, $A_{2a}AR$, $A_{2b}AR$, and A_3AR have been identified. We have detected a rapid increase in the plasma adenosine level during the incidence of adjuvant-induced arthritis in rats [14]. Plasma adenosine level correlated with the severity of arthritis in rats. We have also shown that MTX-induced suppression of osteoclastogenesis is canceled by the administration of adenosine through $A_{2b}AR$, the low affinity AR and that MTX suppressed osteoclastogenesis by suppressing the level of RANKL without affecting OPG production [14].

The level of OPG expression was suppressed without affecting the level of RANKL expression when MTX-suppressed cultures were treated with adenosine, which resulted in the increase in the ratio of RANKL/OPG and augment in osteoclastogenesis.

MTX-treatment seems to have some pivotal relationship with the metabolism of nucleic acids, however, target molecules concerning the MTX-induced suppression of inflammatory bone destruction remains unclear. In our previous study, we have found that deoxyadenosine (dAdo) is the only deoxynucleoside which clearly canceled MTX-induced suppression of osteoclastogenesis [14].

As dAdo is known to regulate proliferation of immune cells and partly share metabolic pathways with adenosine. Adenosine deaminase (ADA) is a key enzyme which metabolize both of adenosine and deoxyadenosine to produce inosine and deoxyinosine perpetually. Deficiency in ADA activity results in a marked elevation of intracellular level of these nucleosides to cause death of immune cells, which leads to the incidence of immune deficiency.

Here we investigated on a possible involvement of ADA in the regulation of MTX- induced suppression of osteoclastogenesis in rats with adjuvant-induced arthritis (arthritic rats). We also examined a possible regulatory role of dAdo, another substrate for ADA, in the MTX-induced suppression of inflammatory bone destruction.

We have obtained lines of evidence suggesting that ADA is one of the target molecules for MTX. MTX-induced suppression of inflammatory bone destruction observed in arthritic rats was clearly canceled by the administration of dAdo.

Materials and Methods

Materials

α-Minimum Essential Medium (αMEM; cat#12000-022) and penicillin-streptomycin were purchased from Invitrogen (Grand Island, NY). RANKL was purchased from Peprotech EC (London, UK). 1α, 25-dihydroxy vitamin D_3 (1α, $25(OH)_2$ D_3) was from Biomol Research Laboratories (Plymouth Meeting, UK). MTX, deoxyadenosine, caffeine and MRS1754 were obtained from Sigma (St.Louis, MO). Heat-killed *Mycobacterium butyricum* and mineral oil were obtained from Difco Laboratories (Detroit, MI). Anti-adenosine deaminase antibody (H-300) was obtained from Santa Cruz Biotechnology (Santa Cruz, CA). Second antibody: Alexa Fluor® 488 Goat

Anti-Rabbit IgG (H+L) for immunofluorescence was obtained from Life Technologies Corporation (Tokyo, Japan). Peroxidase-conjugated anti-rabbit IgG antibody and enhanced chemiluminescence (ECL) kits were purchased from Amersham Biosciences (Buckinghamshire,UK)

Animal Experiments and Ethics Committee Approval

Sprague Dawley (SD) rats and Lewis rats were obtained from Kyudo Co. Ltd. (Tosu, Japan). All animal experiments were performed according to the guide-lines for "Care and Use of Animals of Kyushu University.

Bone Marrow Culture System

Formation of osteoclast-like multinucleated cells (MNCs) from rat bone-marrow cells was performed as described previously [15, 16]. Briefly, bone-marrow cells were obtained from the tibia and femur of 4-6 weeks old male SD rats. Cells were seeded into 24-well culture plates (10^6 cells/well) and cultured in αMEM containing 15% FBS in the presence of 10^{-8}M 1α, 25$(OH)_2$ D $_3$ and 10% (V/V) heat-treated ROS17/2.8 cell conditioned medium (ht-ROS CM). After 4 or 5 days of culture in the presence of various concentrations of MTX, cells were fixed with acetone-citrate-formaldehyde, and stained for TRAP using a Leukocyte acid phosphatase kit. TRAP-positive cells containing more than three nuclei were counted as osteoclast-like multinucleated cells (MNCs). Osteoclast were also detected by staining with osteoclast-specific anti-Kat1-antigen monoclonal antibody [17, 18]. For stromal cell-free cultures, bone marrow stromal cells were depleted by Sephadex G-10 (GE Healthcare, Uppsala, Sweden) column as described previously [15]. Non-adherent bone marrow cells were cultured in the presence of 10^{-8} M 1 α, 25$(OH)_2$ D $_3$, 20 ng/ml sRANKL and 10% (V/V) ht ROSCM. Cells were cultured in 96-well culture plates (3 ×10^5cells/well) for 4 days. Cells were fixed and stained for TRAP.

Induction of Adjuvant Arthritis in Lewis Rats and μCT Analysis

Rats with adjuvant-induced arthritis was prepared as described previously [14, 18]. Briefly, 5 weeks-old female Lewis rats were intradermally injected at the base of the tail with complete Freund adjuvant (CFA) consisting of 25mg/kg heat-killed *M. butyricum* (Difco Laboratories, Detroit, MI) suspended in mineral oil. In control experiments, rats were injected with mineral oil alone. All animals were anesthetized with diethyl ether or isoflurane prior to adjuvant injections. MTX (1mg/kg/week) or vehicle (phosphate buffered saline, PBS) was injected intraperitoneally. These injections were performed once a week beginning at 3 days after the adjuvant injection. Deoxyadenosine or vehicle (PBS) was injected into the ankle-joint cavities of CFA-injected rats. Thirty-gauge needles were used to inject 80 μg dAdo (in 30 μL PBS) around the articular cavity of the left ankle joint. As an internal control, 30 μL PBS was injected around the right ankle joint. Injections were performed every 3 days beginning on the same day as MTX-injections (3 days after the adjuvant injection). The rats were sacrificed on day 21 after the first injection of adjuvant of the same rat. The level of bone destruction of hind paws was imaged by using a μCT analysis system (SKYSCAN-1076KHS Inc., SkyScans, Kontich, Belgium) and SKYSCAN software.

Histological Analysis

After fixation by perfusion with 4% ($^{W}/_{V}$) paraformaldehyde/PBS, tissue blocks were taken from the hind paws (tarsal bones and tibia) of Lewis rats and immersed in the same fixative for overnight at 4 °C, followed by washing in PBS overnight at 4 °C. After decalcification in 10% (W/V) ethylene diamine tetraacetate (EDTA) for 3 weeks at 4 °C and dehydration in ethanol series, the tissue blocks were embedded in paraffin. Sections (6μm) of ankle joints including tibia (*Ti*) and talus (*Ta*) were also stained for TRAP or hematoxylin-eosin (HE).

Semi-quantitative RT-PCR

Total cellular RNA was extracted from cultured cells using TRIzol Reagent (Invitrogen, Carlsbad, CA, USA) and subjected to semi-quantitative RT-PCR using RT-PCR kit (Takara, Kyoto, Japan) according to the manufacturer's protocol. The primers used for PCR were as follows: rat adenosine deaminase forward, 5'-TCC TGG CCA AGT TCG ATT AC-3'; rat adenosine deaminase reverse, 5' - GCC CTG GTT TAC AAG ATC CA-3';

rat adenosine A_1 receptor forward, 5'- CAT TGG GCC ACA GAC CTA CT-3'; rat adenosine A_1 receptor reverse, 5'- CAA GGG AGA GAA TCC AGC AG-3'; rat adenosine A_{2a} receptor forward, 5'-TAC CTG GCC ATC ATT CTC TC-3'; rat adenosine A_{2a} receptor reverse, 5'-TGG GTA GGA AGC TCC ACA TC-3'; rat adenosine A_{2b} receptor forward, 5'-CTT CTG CAC GGA CTT TCA CA-3'; rat adenosine A_{2b} receptor reverse, 5'- GGT GGC ACG GTC TTT ACT GT-3'; rat adenosine A_3 receptor forward, 5'- CGG CTT GGA TTA CAT GGT CT-3'; rat adenosine A_3 receptor reverse, 5'-CAG CAC AAG GCA AAC AAG AA-3'; rat RANKL forward, 5'-CGT TTG CTC ACC TCA CAA TC-3'; rat RANKL reverse, 5'-ACG CTA ATT TCC TCA CCA GC-3'; rat OPG forward, 5'-ATT GGC TGA GTG TTC TGG TG-3'; rat OPG reverse, 5'-GCC CAG TGA CCA TTC CTA AC-3'; rat NFATc1 forward, 5'- CAA CGC CCT GAC CAC CGA TAG-3'; rat NFATc1 reverse rat, 5'-GGC TGC CTT CCG TCT CAT AGT-3' ; rat ADA forward, 5'-TCCTGGCCAAGTTCGATTAC-3', rat reverse, 5'-GCCCTGGTTTACAA-GATC CA-3'; GAPDH forward, 5'-GGT GAT GCT GGT GCT GAG TA-3'; rat GAPDH reverse, 5'-ACT GTG GTC ATG AGC CCT TC-3'. The PCR products were subjected to electrophoresis on 2% agarose gels and visualized by ethidium bromide staining with UV light illumination.

Western Blotting

Areas of ankle joint containing distal tibia and talus were dissected from Lewis rats and were frozen immediately in liquid nitrogen. Frozen tissues were ground into powder under liquid nitrogen using small mortar and pestle. Then the total proteins were extracted in lysis buffer containing 20 mM Tris-HCl (pH7.2), 10 mM EDTA, 0.3 M NaCl, 0.1% Triton X-100, 0.05%Tween 20 and protease inhibitor cocktail (Sigma) overnight at 4 °C. The lysates, containing equal amounts of proteins, were applied on 13% SDS–PAGE followed by transfer to nitrocellulose membranes. After blocking with 5% nonfat dry milk, the membranes were probed with anti-ADA rabbit polyclonal antibody and then reacted with peroxidase-conjugated anti-rabbit IgG antibody followed by detection with ECL kit. The Same membranes were stripped and reprobed with anti-GAPDH.

Immunofluorescence Staining

Deparaffinized sections (6μm) were blocked with 10% normal goat serum for 1 hr at room temperature. These sections were incubated with anti-ADA rabbit polyclonal antibody (H-300) diluted with 1% normal goat serum (1:200 dilution) in a humidified chamber for overnight at 4°C. After washing with PBS, sections were incubated with goat anti-rabbit IgG (H+L) (A11034) conjugated with Alexa Fluor® 488 diluted with 1% normal goat serum (1:500) for 1 h at room temperature. After rinsing, observation was performed by a fluorescence microscope (Carl Zeiss; Axiovert 100, Germany).

Statistical Analysis

All data are presented as mean ± SEM. Statistical analysis was performed using Student's *t*-test. Data shown are representative results from three independent experiments.

Results

Deoxyadenosine Canceled MTX-induced Suppression of Osteoclastogenesis

In rat whole bone marrow cultures, formation of TRAP-positive MNCs was markedly suppressed by MTX. This suppression was completely abolished by the addition of dAdo in a dose-dependent manner, as shown in (Figure 1a and b). MTX markedly inhibited the formation of MNCs expressing Kat1-antigen, a specific marker for rat osteoclasts, and addition of dAd completely abolished its suppression in the formation of Kat1-antigen-positive osteoclasts (Figure 1c).

When bone marrow cells were cultured in the absence of MTX, addition of dAdo did not affect osteoclastogenesis (Figure 1d). When stromal cell-free non-adherent bone marrow cells (NABMCs) were utilized, neither MTX-induced suppression nor dAdo-mediated recovery of osteoclastogenesis was observed (Figure 1e). These data suggest that MTX-induced suppression and cancellation by deoxyadenosine in osteoclastogenesis was mediated through marrow stromal cells.

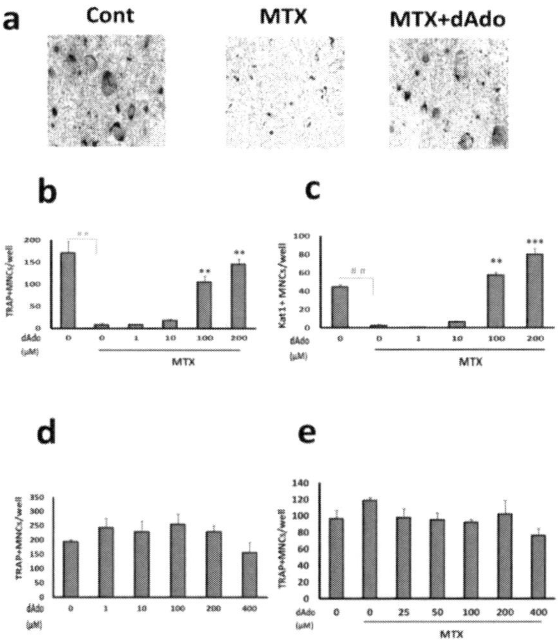

Figure 1. Deoxyadenosine canceled MTX-induced suppression of osteoclastogenesis. (a-c) Abolishment of MTX-induced suppression of osteoclastogenesis by dAdo. Bone-marrow cells were cultured for forming osteoclasts in 24-well culture plates (10^6 cells/well) as described in MATERIALS AND METHODS with or without 1μM of MTX in the presence of various concentrations of dAdo. After 4 days of culture, the cells were stained for TRAP (a,b) or for Kat1-antigen (c) as described in MATERIALS AND METHODS. (a) Microscopic observation of the TRAP-positive MNCs. Cont: control in the absence of MTX, MTX: in the presence of 1μM MTX, MTX+dAdo: in the presence of 1 μM MTX and 200 μM dAdo. (b,c) Abolishment of MTX-induced suppression of osteoclastogenesis by dAdo in a dose-dependent manner. Number of TRAP-positive MNCs were counted (b) and that of Kat1-positive MNCs were counted (c). Data present means±SEM from four cultures (n=4). Data were analyzed by Student's *t*-test * $p<0.05$, ** $p<0.01$, *** $p<0.001$ if compared with the control culture in the absence of MTX. # $p<0.05$, ## $p<0.01$ if compared among indicated pairs. (d) Effect of dAdo on osteoclastogenesis in the absence of MTX. (e) MTX–induced suppression of osteoclastogenesis was dependent on bone marrow stromal cells. For stromal cell-free cultures, bone marrow stromal cells were depleted by Sephadex G-10 (GE Healthcare, Uppsala, Sweden) column as described in MATERIALS AND METHODS. The stromal cell-free bone marrow cells were cultured in 96-well culture plates (3×10^5cells/well) in the presence of various concentrations of dAdo with or without 1μM MTX. After 4 days of culture, the cells were fixed and stained for TRAP. Data present means±SEM from four cultures. Data were analyzed by Student's *t*-test * $p<0.05$, ** $p<0.01$, *** $p<0.001$ if compared with the control culture in the absence of MTX.

Recovery Effect of dAdo was Partially Prevented by Caffeine, Antagonist for High Affinity Adenosine Receptors

To investigate a possibility that ARs are involved in the abolishment of MTX-induced suppression of osteoclastogenesis by dAdo, we assessed the effects of adenosine receptor antagonists (Figure 2). MTX-induced suppression of osteoclastogenesis was partially blocked by the addition of caffeine, an antagonist for the high affinity adenosine receptors, A_1AR, $A_{2a}AR$ and A_3AR (Figure 2a).

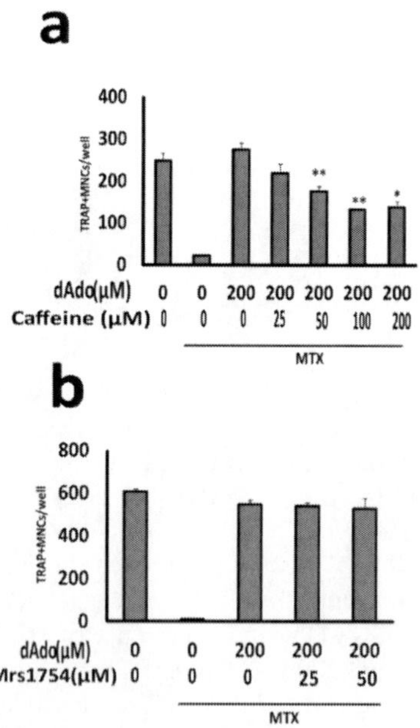

Figure 2. The recovery effect of dAdo was partially blocked by caffeine but not by MRS1754. Bone-marrow cells were cultured for forming osteoclasts in 24-well culture plates (10^6cells/well) with or without MTX in the present or absence of 200μM dAdo. These cultures were treated with various concentration of caffeine (a) or MRS1754 (b). After 4 days of culture, cells were fixed and stained for TRAP and TRAP-positive MNCs were counted. Data present means±SEM from four cultures. Experiments were performed 3 times and the representative data were shown. Data were analyzed by Student's *t*-test. $^{\#\#\#}p<0.001$ if compare with the control culture in the absence of MTX. $^{*}P<0.5$, $^{**}P<0.01$, $^{***}P<0.001$ if compared with culture treated with dAdo in the absence of caffeine (for (a)) or MRS1754 (for (b)).

However, as shown in (Figure 2b), MRS1754, the selective antagonist for the low affinity receptor $A_{2b}AR$ did not affect this canceling effect of dAdo from MTX-induced suppression of osteoclastogenesis (Figure 2b). These data suggested that dAdo-mediated abolishment of MTX suppression is partly mediated through high affinity adenosine receptors, A_1AR, A_{2a} AR or A_3AR.

Deoxyadenosine Significantly Abolished the Therapeutic Effects of MTX on Inflammatory Bone Destruction in Arthritic Rats

Arthritic rats were sacrificed on day 21 after the first injection of adjuvant with or without MTX- treatment. The level of bone destruction in the hind paws injected with or without dAdo was imaged by μCT analysis. As shown in (Figure 3a), MTX markedly suppressed inflammatory bone destruction in arthritic rats. Injection of dAdo around the ankle-joint cavities of arthritic rats completely diminished the therapeutic action of MTX on inflammatory bone destruction (Figure 3a). TRAP-staining data clearly demonstrated that MTX markedly suppressed osteoclastogenesis in the distal tibia and talus of arthritic rats and that dAdo-treatment completely abolished the suppressive effect of MTX in osteoclastogenesis *in vivo*. Quantitative μCT analysis of the talus was shown concerning four parameters: percent bone volume (BV/TV) (%), trabecular thickness (Tb.Th) (mm), trabecular number (Tb.N/mm) and trabecular separation (Tb.Sp) (mm). As shown in Figure 3b, μCT analysis clearly showed that the marked bone loses observed in arthritic rat was significantly suppressed by MTX-treatment and that injection of dAdo around the articular joints completely cancelled MTX-induced suppression.

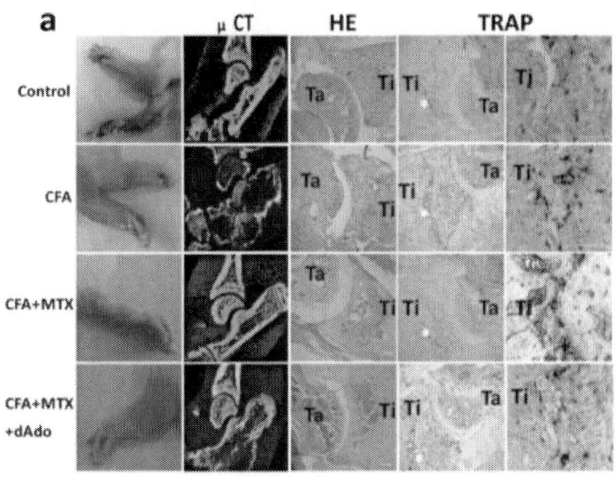

Figure 3. (Continued).

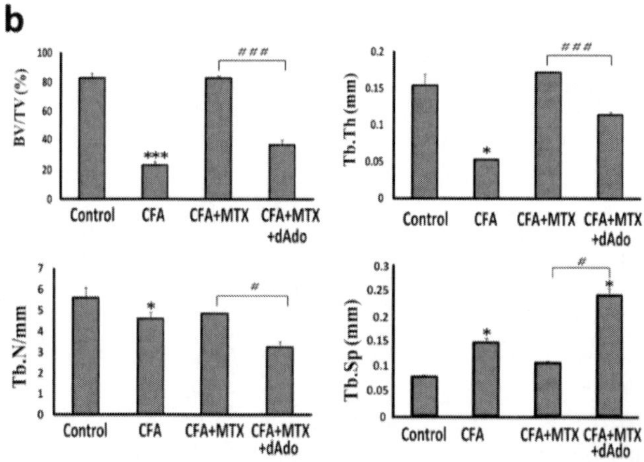

Figure 3. Deoxyadenosine significantly abolished the therapeutic effects of MTX on inflammatory bone destruction in rats with adjuvant-induced arthritis.
(a) MicroCT analysis and histochemical analysis: Arthritic rats were induced by the injection of complete Freund's adjuvant (CFA) as described in MATERIALS AND METHODS. Arthritic rats were treated with MTX (1mg/kg/week) (CFA+MTX) or MTX plus dAdo (CFA+MTX+ dAdo). As the control, rats were injected mineral oil alone (Control). After analyzing the ankle joints by use of μCT (μCT), bone tissues were processed for histological analysis. Paraffin sections (6μm) were prepared and stained for hematoxylin eosin (HE) or for TRAP activity (TRAP). (In TRAP staining data, the right panels are the high magnification view of the left panel). Ti: tibia, Ta: talus. Bars 50 μm. (b) Quantitative demonstration of μCT analysis: The talus was analyzed by μCT concerning the four parameters: Percent bone volume (BV/TV) (%), trabecular thickness (Tb.Th)(mm), trabecular number (Tb.N)/mm and trabecular separation (Tb.Sp)(mm). Data present mean±SEM for four rats. *$p<0.05$, ***$p<0.001$ if compared with control (Control). #$P<0.05$, ###$P<0.001$ if compared among indicated pairs.

Suppression of RANKL Expression Mediated by MTX Was Cancelled by dAdo

In bone marrow cultures for evaluating osteoclastogenesis, MTX-induced suppression of osteoclastogenesis was canceled by the addition of dAdo (Figure 1a). In the same condition, the total RNA was extracted and analyzed by semi-quantitative RT-PCR (Figure 4). MTX did not affect expression of OPG mRNA expression but markedly suppressed expression of RANKL and NFATc1. Addition of dAdo almost completely recovered expression of RANKL and NFATc1. Furthermore, dAdo treatment inhibited expression of OPG mRNA. Therefore dAdo-treatment markedly increased the expression-ratio of RANKL/OPG (Figure 4a). We have examined the expression of ARs. Although MTX did not affect the expression of mRNA for A_1AR and A_{2b}AR, MTX augmented expression of mRNA for A_{2a}AR and slightly increased expression of A_3AR (Figure 4b). The level of mRNA expression of ARs were not affected by the administration of dAdo.

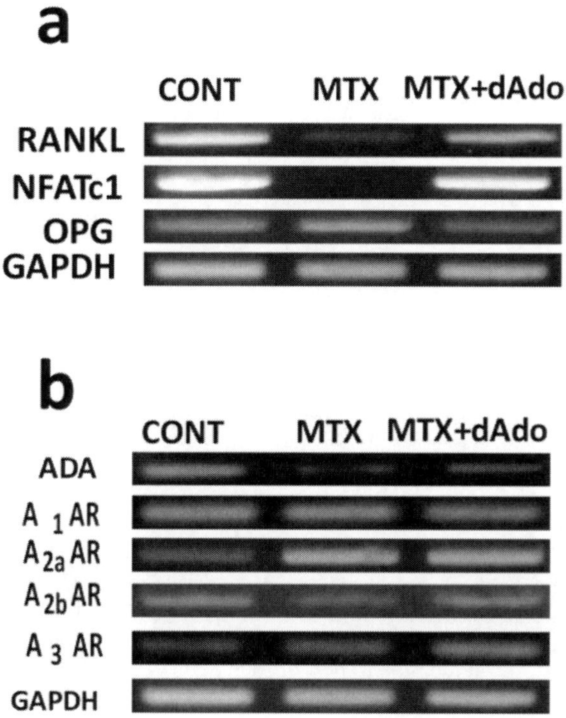

Figure 4. Semi-quantitative RT-PCR analysis of gene expression concerning osteoclastogenesis and adenosine metabolism. Bone marrow cells were cultured for forming osteoclasts as described in MATERIALS AND METHODS with or without 1μM MTX (+MTX) in the presence or absence of 200μM dAdo. Total RNAs were extracted and semi-quantitative RT-PCR was performed using primer sets for RANKL, NFATc1, OPG, GAPDH (a) or those for ADA, A_1AR, A_{2a}AR, A_{2b}AR, A_3AR and GAPDH (b). PCR products were analyzed by 2% agarose gels.

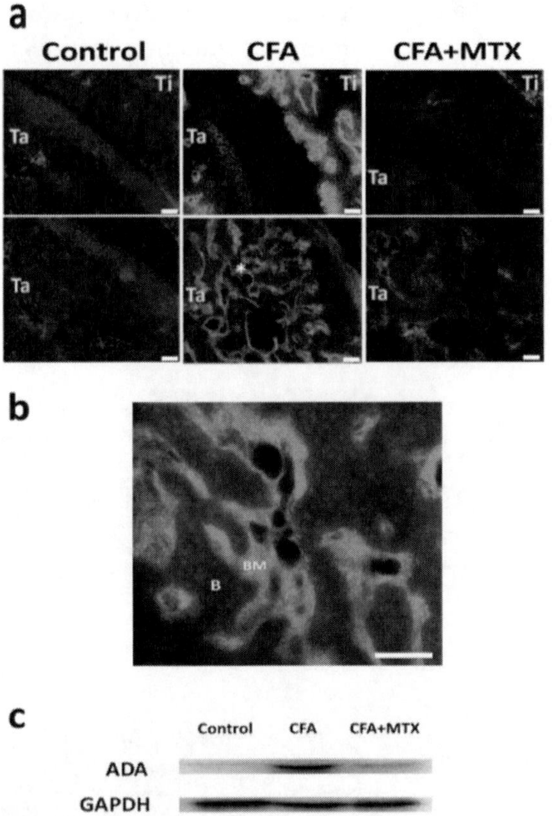

Figure 5. MTX inhibits ADA expression in bone destruction sites of arthritic rats. (a) Immunohistochemical analysis. Hind paw-joint sections of control rats (Control), arthritic rats (CFA), arthritic rats injected with MTX (CFA+MTX) were incubated with anti-ADA polyclonal antibody (rabbit) followed by detection with anti-rabbit IgG conjugated with AlexaFluor 488 as described in MATERIALS AND METHODS. Upper panels are the immunofluorescence observation of the ankle joint. Lower panels are the immunofluorescence view of talus. Ti: tibia, Ta: talus. Bars: 50μm. (b) Detection of intense ADA immunoreativity in bone marrow area in arthritic rats. The high magnification view of the asterisk area of the lower central panel of (a). B:bone, BM: bone marrow. Bars 50 μm. (c) Western blot analysis. Protein samples of the area of the ankle joints containing distal tibia and talus were prepared from control rats (Control), arthritic rats (CFA), arthritic rats injected with MTX (CFA+MTX) as described in MATERIALS AND METHODS. The cell lysates containing equal amounts of proteins were subjected to 13% SDS-PAGE followed by immunoblotting as described in MATERIALS AND METHODS. Membrane sheets were reacted with anti-ADA antibody (ADA) or anti-GAPDH antibody (GAPDH).

MTX Suppressed Expression of ADA in Adjuvant-Induced Arthritis

Our previous report showed that MTX markedly suppressed osteoclastogenesis and adenosine cancels its suppression [14]. As ADA is a key enzyme regulating the intracellular level of adenosine and dAdo, we have examined the expression of ADA in MTX-induced suppression of bone destruction. To know whether ADA is involved in the mechanisms of MTX-mediated inhibition of inflammatory bone destruction, we have investigated the expression of ADA in the joint tissue of the arthritic rats by use of immunofluorescence methods.

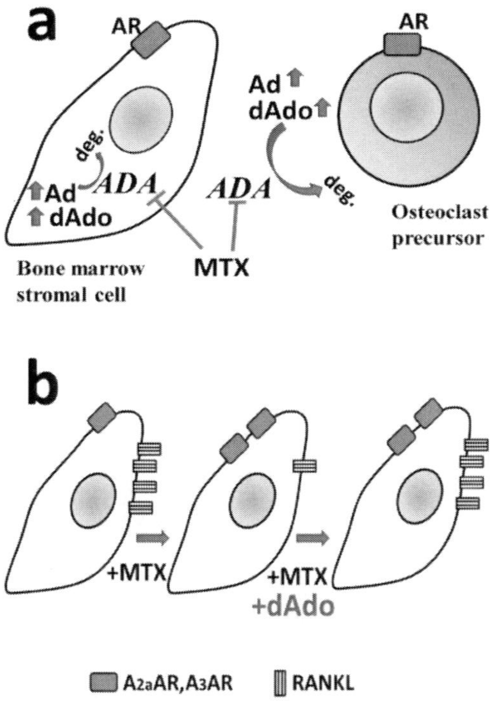

Figure 6. A possible Involvement of ADA and dAdo in the regulation of MTX-induced suppression of osteoclastogenesis: Illustration diagram. (a) Inhibition of ADA-expression by MTX. (b) MTX-induced suppression of RANKL- expression was canceled by dAdo.

As shown in Figure 5, immunoreactivity of ADA was quite intense in bone destruction area of arthritic rats, while control rats showed minimum level of ADA expression. High magnification view of ADA-positive area of arthritic rats showed an intense immunoreactivity in bone marrow area. A diffuse staining pattern was observed in the bone marrow area, in which not only bone marrow cells but also extra-cellular spaces were stained. MTX-treatment markedly reduced the staining level with anti-ADA antibody in the corresponding area. Western blotting analysis confirmed a clear induction of ADA protein in arthritic rats and a marked suppression of ADA by MTX (Figure 5c). Semi-quantitative RT-PCR using bone marrow cultures for evaluating osteoclastogenesis showed that the expression of ADA mRNA was detected in cultures for evaluating osteoclastogenesis and it was markedly suppressed by MTX treatment. MTX-induced suppression of ADA gene expression was slightly canceled by the addition of dAdo, although its expression level was still significantly suppressed.

Discussion

In the current study, we have demonstrated a marked induction of ADA in the bone destruction sites of arthritic rats. As a diffuse immunoreactivity of ADA was observed in the bone marrow area, ADA seems to be localized not only in the producer cells but also in the extracellular spaces of the bone marrow cavity in arthritic rats. Nakamachi et al. [19] reported the specific increase in the enzyme activity of ADA in rheumatoid synovial fibroblasts. Our finding concerning a high level of protein in bone marrow cavity is supposed to be related to a direct function of ADA in bone metabolism [20]. MTX-treatment markedly suppressed expression of ADA in the current study, which was also confirmed by immunobloting. Our findings clearly demonstrated an involvement of ADA in the regulation of inflammatory bone destruction. MTX is reported to increase extracellular adenosine concentration [6, 21]. Suppression of ADA expression by MTX is supposed to result in a marked accumulation of adenosine and dAdo. In this condition, exogenously administered dAdo is supposed to be markedly accumulated without degradation as the level of ADA is maintained in a very low level by MTX. In patients with ADA-deficiency combined with severe immune-deficiency disease, it has been reported that adenine deoxynucleosides and deoxynucleotides are overproduced in these patients [22]. Erer et al. [23]

reported that serum ADA levels in RA patients are found to be higher than healthy controls. Their clinical observations are well consistent with findings in the current study in which we have detected abundant ADA in the bone destruction sites of arthritic rats.

By use of marrow culture system for forming osteoclasts, we have shown that MTX-induced suppression of RANKL expression was clearly canceled by the addition of dAdo. In this condition, expression of OPG was slightly suppressed by the addition of dAdo. These observations strongly suggest that dAdo markedly augment in the ratio of RANKL/OPG, which allows abundant osteoclastogenesis in the presence of MTX (Figure 6). In MTX-suppressed conditions, dAdo-induced induction of osteoclastogenesis was partly blocked by the addition of caffein, an antagonist for the high affinity adenosine receptors, A_1AR, $A_{2a}AR$ and A_3AR, but not blocked by the addition of antagonists for the low affinity receptor, $A_{2b}AR$. Among these receptors, MTX-treatment stimulated $A_{2a}AR$ and A_3AR and further treatment with dAdo did not change their expression level. It has been so far unclear that dAdo binds to ARs, our current research could predict a possibility that dAdo utilizes high affinity ARs in some conditions. Kara et al. [24] demonstrate that A_1AR is required for normal and pathological osteoclastogenesis. Ochaion et al. [25] reported a highly selective agonist for A_3AR effectively suppressed inflammation and bone destruction accompanying arthritis. Rath-Wolfson et al. [26] reported on an involvement of A_3AR in bone resorption in rats with adjuvant-induced arthritis. These data combined with our findings suggest that high affinity receptors for adenosine are involved in the dAdo-mediated osteoclastogenesis. In contrast, Mediero et al. [27] recently reported that $A_{2a}AR$ agonist, CGS21680, inhibited osteoclast differentiation and function. Femurs of $A_{2a}AR$ gene deficient mice show a significantly decreased bone formation with an increased TRAP-positive osteoclasts, strongly suggesting that $A_{2a}AR$ acts as a negative regulator for osteoclastogenesis. Thus ARs could act as the positive or negative regulators in osteoclastogenesis [28].

Among nucleotides, adenine-bound nucleotides show unique regulatory activities for osteoclastogenesis as well as for bone resorption. Pellegatti et al. [29] has shown that P2X7, the receptor for ATP, is involved in osteoclastogenesis. Receptors for ADP, P2Y1 and P2RY12, are also demonstrated to be involved in the regulation of osteoclastogenesis [30, 31]. The regulatory system using adenine-bound nucleosides and nucleotides in osteoclastogenesis might have been evolutionally developed in vertebrate. In general, extracellular ATP is utilized as a neurotransmitter in neuro-synapse and it has also been shown that ATP secreted from apoptotic cells acts as a "find-me signal" for phagocytes through $P2Y_2$ ATP receptors expressed on macrophages [32].

Our previous report showed MTX-induced suppression of osteoclastogenesis was canceled by the addition of adenosine through the low affinity adenosine receptor, $A_{2b}AR$ [14]. In this case, MTX-induced suppression of RANKL expression was not canceled by adenosine but adenosine-treatment markedly suppressed expression of OPG, which finally resulted in an increase in the ratio of RANKL/OPG and subsequent augment in osteoclastogenesis. Major cause of the difference in the influence on RANKL expression between adenosine and dAdo could be attributed to a difference in the receptors utilized, although cell surface specific receptors for dAdo has not so far been identified. Extracellular dAdo could also been incorporated into cells through nucleoside transporters [33]. Details remains to be clarified concerning the sensing mechanism of osteoclast precursors to extracellular dAdo. In general, intracellular pool of deoxynucleosides is supplied by *de novo* synthesis of deoxynucleosides or uptake of extracellular deoxy-nucleosides through nucleoside transporter. Major source of extracellular deoxynucleosides is DNA from apoptotic cells [34]. For normal cell growth, well balanced deoxynucleoside pool is essential. When pool of some deoxynucleoside lapse into beeing insufficient (hypo-conditions) or extra-sufficient (hyper-conditions), "replication stress" occurs and DNA synthesis is inhibited. Replication stress has recently shown to be a potent driver of functional decline in aging hematopoietic stem cells [35]. Austin et al. [36] have demonstrated that the deoxycytidine kinase plays a critical role to avoid "replication stress" in hematopoietic cells. This enzyme phosphorylates deoxycytidine to form dCMP. It is utilized as a precursor of dCTP in the process of DNA synthesis. Deoxycytidine kinase is able to catalyze dAdo and deoxyguanosine to convert into dAMP and dGMP, respectively. Deficiency in this enzyme severely affects differentiation of hematopoietic cells. In our current study, dAdo-mediated stimulation of osteoclastogenesis in the presence of MTX could explain by the action of deoxycytidine kinase to metabolize dAdo. In general, a fraction of the cellular pool of ribonucleosides is converted into deoxyribonucleosides by the ribonucleoside reductase [37]. Prompt phosphorylation of dAdo formed from adenosine by ribonucleoside reductase would also contribute to avoid "replication stress". Involvement of dAdo in the regulation of osteoclastogenesis could be related to the regulation of "replication stress". Further investigations are required to elucidate the mechanism of induction of osteoclastogenesis by dAdo-treatment.

Acknowledgments

We thank Prof. Tang of the second hospital in He Bei Medical University, China for encouragements.

Disclosure/conflict of interest: The authors declare no conflict of interest.

References

[1] Teitelbaum SL. Bone resorption by osteoclasts. *Science* 2000; 289(5484):1504-1508.
[2] Boyle WJ, Simonet WS, Lacey DL. Osteoclast differentiation and activation. *Nature* 2003; 423(6937):337-342.
[3] Takayanagi H. Osteoimmunology: shared mechanisms and crosstalk between the immune and bone systems. *Nat Rev Immunol* 2007; 7(4):292-304.
[4] Lacey DL, Timms E, Tan HL, Kelley MJ, Dunstan CR, Burgess T, et al. Osteoprotegerin ligand is a cytokine that regulates osteoclast differentiation and activation. *Cell* 1998; 93(2):165-176.
[5] Farber S, Diamond LK. Temporary remissions in acute leukemia in children produced by folic acid antagonist, 4-aminopteroyl-glutamic acid. *N Engl J Med* 1948; 238(23):787-793.
[6] Cronstein BN. Low-dose methotrexate: a mainstay in the treatment of rheumatoid arthritis. *Pharmacol Rev* 2005; 57(2):163-172.
[7] Thompson PA, Murry DJ, Rosner GL, Lunagomez S, Blaney SM, Berg SL, et al. Methotrexate pharmacokinetics in infants with acute lymphoblastic leukemia. *Cancer Chemother Pharmacol* 2007; 59(6):847-853.
[8] Weinblatt ME, Coblyn JS, Fox DA, Fraser PA, Holdsworth DE, Glass DN, et al. Efficacy of low-dose methotrexate in rheumatoid arthritis. *N Engl J Med* 1985; 312(13):818-822.
[9] Buckley LM, Leib ES, Cartularo KS, Vacek PM, Cooper SM. Effects of low dose methotrexate on the bone mineral density of patients with rheumatoid arthritis. *J Rheumatol* 1997; 24(8):1489-1494.

[10] Choi HK, Seeger JD, Kuntz KM. A cost-effectiveness analysis of treatment options for patients with methotrexate-resistant rheumatoid arthritis. *Arthritis Rheum* 2000; 43(10):2316-2327.

[11] Barbieri M, Wong JB, Drummond M. The cost effectiveness of infliximab for severe treatment-resistant rheumatoid arthritis in the UK. *Pharmacoeconomics* 2005; 23(6):607-618.

[12] Jacobson KA, Gao ZG Adenosine receptors as therapeutic targets. *Nature Rev. Drug Discov.* 2006; 5:247-264.

[13] Hasko G, Cronstein BN. Adenosine: an endogenous regulator of innate immunity. *Trends Immunol* 2004; 25(1):33-39.

[14] Teramachi J, Kukita A, Li YJ, Ushijima Y, Ohkuma H, Wada N, et al. Adenosine abolishes MTX-induced suppression of osteoclastogenesis and inflammatory bone destruction in adjuvant-induced arthritis. *Lab Invest*; 2011; 91(5):719-731.

[15] Kukita A, Kukita T, Shin JH, Kohashi O. Induction of mononuclear precursor cells with osteoclastic phenotypes in a rat bone marrow culture system depleted of stromal cells. *Biochem Biophys Res Commun* 1993; 196(3):1383-1389.

[16] Kukita A, Kukita T, Hata K, Kurisu K, Kohashi O. Heat-treated osteoblastic cell (ROS17/2.8)-conditioned medium induces the formation of osteoclast-like cells. *Bone Miner* 1993; 23(2):113-127.

[17] Kukita T, Kukita A, Nagata K, Maeda H, Kurisu K, Watanabe T, et al. Novel cell-surface Ag expressed on rat osteoclasts regulating the function of the calcitonin receptor. *J Immunol* 1994; 153(11):5265-5273.

[18] Li YJ, Kukita A, Watanabe T, Takano T, Qu P, Sanematsu K, et al. Nordihydroguaiaretic acid inhibition of NFATc1 suppresses osteoclastogenesis and arthritis bone destruction in rats. *Lab Invest* 2013; 92(12):1777-1787.

[19] Nakamachi Y, Koshiba M, Nakazawa T, Hatachi S, Saura R, Kurosaka M, Kusaka H, Kumagai S. Specific increase in enzymatic activity of adenosine deaminase 1 in rheumatoid synovial fibroblasts. *Arth. Rheum.* 2003; 48(3)668-674.

[20] Yu S, Sharma R, Nie D, Jiao H, Im HJ, Lai Y, et al. ADAR1 ablation decreases bone mass by impairing osteoblast function in mice. *Gene* 2013; 513(1): 101-110.

[21] Riksen NP, Barrera P, van den Broek PH, van Riel PL, Smits P, Rongen GA. Methotrexate modulates the kinetics of adenosine in humans in vivo. *Ann Rheum Dis* 2006; 65(4):465-470.

[22] Donofrio J, Coleman MS, Hutton JJ, Daoud A, Lampkin B, Dyminski J. Overproduction of adenine deoxynucleosides and deoxynucleotides in adenosine deaminase deficiency with severe combined immunodeficiency disease. *J Clin Invest* 1978; 62(4):884-887.

[23] Erer B, Yilmaz G, Yilmaz FM, Koklu S. Assessment of adenosine deaminase levels in rheumatoid arthritis patients receiving anti-TNF-α therapy. *Rheumatol Int* 2009; 29:651-654.

[24] Kara FM, Chitu V, Sloane J, Axelrod M, Fredholm BB, Stanley ER, et al. Adenosine A1 receptors (A1Rs) play a critical role in osteoclast formation and function. *FASEB J* 2010 ; 24(7):2325-2333.

[25] Ochaion A, Bar-Yehuda S, Cohn S, Del Valle L, Perez-Liz G, Madi L, et al. Methotrexate enhances the anti-inflammatory effect of CF101 via up-regulation of the A3 adenosine receptor expression. *Arthritis Res Ther* 2006; 8(6):R169.

[26] Rath-Wolfson L, Bar-Yehuda S, Madi L, Ochaion A, Cohen S, Zabutti A, et al. IB-MECA, an A3 adenosine receptor agonist prevents bone resorption in rats with adjuvant induced arthritis. *Clin Exp Rheumatol* 2006; 24(4):400-406.

[27] Mediero A, Kara FM, Wilder T, Cronstein BN. Adenosine A(2A) receptor ligation inhibits osteoclast formation. *Am J Pathol* 2012; 180(2):775-786.

[28] Mediero A, Cronstein BN Adenosine and bone metabolism. *Trends in Endocrinology and Metabolism* 2013; 24(8)290-300.

[29] Pellegatti P, Falzoni S, Donvito G, Lemaire I, Di Virgilio F. P2X7 receptor drives osteoclast fusion by increasing the extracellular adenosine concentration. *FASEB J* 2011; 25(4):1264-1274.

[30] Hoebertz A, Meghji S, Burnstock G, Arnett TR. Extracellular ADP is a powerful osteolytic agent: evidence for signaling through the P2Y(1) receptor on bone cells. *FASEB J* 2001; 15(7):1139-1148.

[31] Su X, Floyd DH, Hughes A, Xiang J, Schneider JG, Uluckan O, et al. The ADP receptor P2RY12 regulates osteoclast function and pathologic bone remodeling. *J Clin Invest* 2012; 122(10):3579-3592.

[32] Elliott MR, Chekeni FB, Trampont PC, Lazarowski ER, Kadl A, Walk SF, et al. Nucleotides released by apoptotic cells act as a find-me signal to promote phagocytic clearance. *Nature* 2009; 461(7261):282-286.

[33] Ham J, Evans BA. An emerging role for adenosine and its receptors in bone homeostasis. *Front Endocrinol (Lausanne) 2012*; 3:113-120.

[34] Arner ES, Eriksson S. Mammalian deoxyribonucleoside kinases. *Pharmacol Ther* 1995; 67(2):155-186.

[35] Flach J, Bakker, Mohrin M, Conroy PC, Pietras, Reynaud D, Alvarez S, Diolaiti ME, Ugarte F, Forsberg EC, Le Beau AA, Stohr BA, Mendez J, Morrison CG, Passegue E. Replication stress is a potent driver of functional decline in aging haematopoietic stem cells. *Nature* 2014; 512:198-202.

[36] Austin WR, Armijo AL, Campbell DO, Singh AS, Hsieh T, Nathanson D, et al. Nucleoside salvage pathway kinases regulate hematopoiesis by linking nucleotide metabolism with replication stress. *J Exp Med* 2012; 209(12):2215-2228.

[37] Fairman JW, Wijerathna SR, Ahmad MF, Xu H, Nakano R, Jha S, et al. Structural basis for allosteric regulation of human ribonucleotide reductase by nucleotide-induced oligomerization. *Nat Struct Mol Biol* 2011; 18(3):316-322.

Index

B

C

D

E

F

G

H

I

J

K

L

M

N

O

Q

R

S

T

U

V

W

Y